# The Beautiful Food Garden

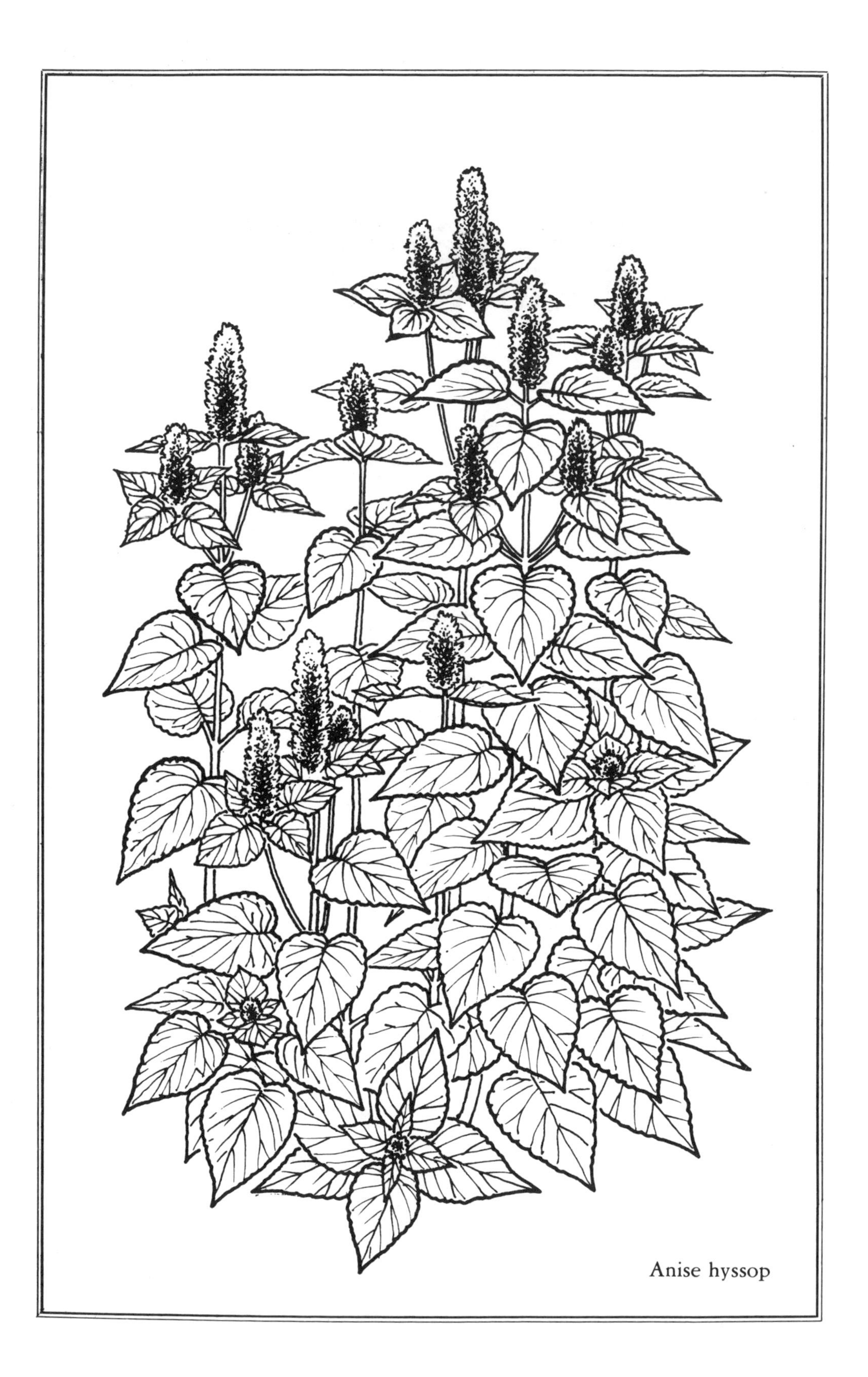

Anise hyssop

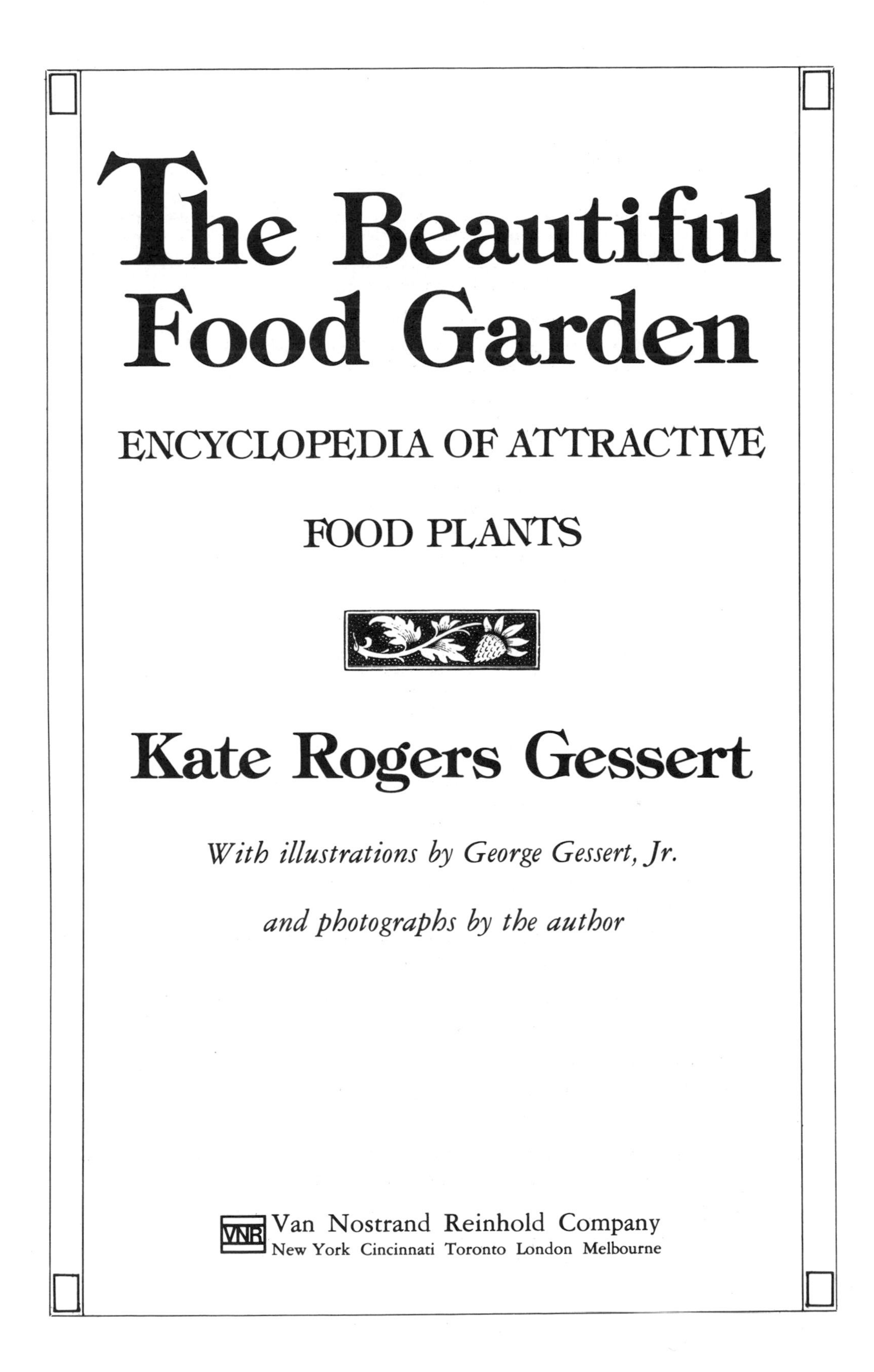

# The Beautiful Food Garden

## ENCYCLOPEDIA OF ATTRACTIVE FOOD PLANTS


**Kate Rogers Gessert**


*With illustrations by George Gessert, Jr.*

*and photographs by the author*


VNR Van Nostrand Reinhold Company
New York Cincinnati Toronto London Melbourne

Library of Congress Catalog Card Number 81-16340
ISBN 0-442-23857-6



Printed in the United States of America

Designed by Ginger Legato

Published by Van Nostrand Reinhold Company Inc.
135 West 50th Street, New York, NY 10020

Van Nostrand Reinhold Publishers
1410 Birchmount Road
Scarborough, Ontario M1P 2E7, Canada

Van Nostrand Reinhold Australia Pty. Ltd.
17 Queen Street
Mitcham, Victoria 3132, Australia

Van Nostrand Reinhold Company Limited
Molly Millars Lane
Wokingham, Berkshire, England

16 15 14 13 12 11 10 9 8 7 6 5 4 3 2 1

**Library of Congress Cataloging in Publication Data**
Gessert, Kate Rogers.
The beautiful food garden.
Bibliography: p. 245
Includes index.
1. Vegetable gardening. 2. Food crops.
3. Landscape gardening. 4. Fruit-culture.
I. Title.
SB321.G4 635 81-16340
ISBN 0-442-23857-6 AACR2

Grateful acknowledgment is made to McGraw-Hill Book Company for permission to reprint an excerpt from *Landscape Architecture: The Shaping of Man's Natural Environment* by John Ormsbee Simonds (New York: McGraw-Hill Book Co., 1961).

Line drawings in Chapters 3 and 4 are by George Gessert, unless otherwise noted.

Other line drawings in Chapter 3, as individually noted, are from *The Vegetable Garden* by Mm. Vilmorin-Andrieux (1885. Reprint. Palo Alto, Ca.: Jeavons-Leler Press, 1976).

Landscape designs in Chapter 2 are by Kate Rogers Gessert, with final drafting by John Sargent, unless otherwise noted.

Color and black-and-white photographs in the book are by Kate Rogers Gessert, unless otherwise noted.

*To my father*

*Major John J. Rogers,*

*whose books were never finished*

# Contents

*Acknowledgments*

*Preface*

1. *Food Plants as Part of the Landscape* 7

2. *Designing the Beautiful Food Garden* 15

3. *Encyclopedia of Attractive Vegetables and Herbs* 43
   *Less Attractive Vegetables and Herbs* 185
   *Plants with Limited Food Uses* 191
   *Possibilities to Explore* 193

4. *Fruits and Nuts for the Beautiful Food Garden* 200

*Appendix: Sources and Information* 229

*Selected Bibliography* 245

*Index* 251

# Acknowledgments

I want to thank not only the people who have helped me directly with this book, but also those who have loved me and believed in me over the years and who have helped me become strong enough to take this on. My deepest gratitude and love go to my husband, George Gessert; my son, Joseph Rogers Gessert; my daughter, Sarah Rogers Gessert; my mother, Margaret Rogers; and my mother-in-law, Mary S. Gessert, who have supported and delighted me while I worked on this book.

I also want to thank Dr. James Baggett for giving me help and space for extensive trial gardens at Oregon State University Vegetable Research Farm and for editing my manuscript; Steve Lindsey for his excellent and enthusiastic landscape architecture teaching; Helen Park for taking care of Joseph, typing the manuscript, and being such a good friend; Esther Park, Helen's daughter, for loving Joseph.

My thanks also to everyone else who took great care of Joseph and freed my mind for work, especially Judy Albrecht and the Gathered School, Mary Starbuck, and Char and Carl Peterson.

Many others have also provided support and knowledge; I wish I had room to mention everybody. I especially want to thank Chuck DeDeurwaerder, Anita Green, Dr. James Green, Deborah Keene, Dr. Harry Mack, Dr. Alfred N. Roberts, Dr. Maxine Thompson, and Dr. Melvin Westwood of Oregon State University; Frances Miner, Debbie Trowbridge, Mollie Rodriguez-Noriega, Daphne Drury, Fred McGourty, Alice Smith, Charles Mazza, Marie Giasi, and many others who taught and helped me at the Brooklyn Botanic Garden; Ann Bettman and Richard Britz of the University of Oregon; Maryanne Caruthers-Akin; Rachel Davis; Jeanne Etter; Megan Hughes; John Sargent; all the panelists who evaluated vegetable varieties at the Vegetable Research Farm; Corvallis gardeners who grew the varieties in their home gardens; and the many seed companies who generously provided seeds for trial at Oregon State University.

My thanks and apologies to the many authors whose information I have incorporated and, for space reasons, could not acknowledge in specific footnotes. These references are listed in the Bibliography.

# Preface

For several years I worked as an instructor at the Brooklyn Botanic Garden. I taught classes for children and teachers, and cared for the plants used in the classes. I also worked with community groups that were planting gardens in vacant lots and with teachers who were growing vegetables in their schools.

For many of the people with whom I worked, plants were a great source of enjoyment and growth. Some people who lived in the city had grown up in the country and missed having plants in their environment. Others who had grown up in cities were not used to plants at all, but they too became deeply interested. Once it began, the love and excitement felt by these strangers to the plant world grew vigorously. During the last few years that I lived there, New York City seemed full of people who were suddenly conscious and fiercely protective of plants.

The community groups with whom I worked often struggled for months to get permission to garden in rubble-strewn vacant lots; they literally created soil out of whatever waste materials they could collect. Some of these community gardeners had trouble deciding whether they should grow food plants or ornamentals; they wanted their gardens to be attractive, and they also wanted to grow food. By coincidence, I was working on a project connected to that question.

During my first year at the Brooklyn Botanic Garden, I had designed a demonstration garden that used herbs and flowers as companion plants to protect vegetables from insects. I could not tell which plant combinations were beneficial to each other, but I was delighted by the garden—by the various colors, textures, and shapes of it and by the feel and fragrance of the herbs, vegetables, and flowers as I moved around it. It seemed to me a much richer, more beautiful and sensuous garden than those with vegetables or flowers alone.

I also took care of an annual flower border in the Botanic Garden Children's Garden, and, during the next several years, I grew vegetables and herbs in beds with the flowers. There were plenty of complications. I wanted to try out attractive new vegetable varieties, but I couldn't always tell from seed catalog descriptions how the plants

Here are seen the contrasting textures of companion plants—nasturtium leaves in the foreground and squash in the background.

were going to look. There were many surprises, both bad and good; my gardens were quite unpredictable.

I left the Brooklyn Botanic Garden after several years to study horticulture at Oregon State University, and several months later my first child was born. For the next few years my husband and I took turns going to school and staying with our son. This sometimes meant passing him quickly from one pair of arms to another in a hallway between classes. Working with Dr. James Baggett at the Oregon State University Vegetable Research Farm, I grew extensive trials of vegetable and herb varieties—five hundred one year and eight hundred the next. I rated these plants in terms of attractiveness, taste, yield, vigor, and insect and disease resistance. Since then I have continued to design and grow gardens using the varieties I have liked best, and other local gardeners have also tried them out.

This book is a distillation of what I have learned. Results of my research on more than one thousand vegetable and herb varieties should help you decide which to try first. My results are not conclusive. Varieties I grew may look different growing in other climates, and new varieties become available every year, but my work can serve as a starting point.

It is time to pull down the artificial barriers that divide our gardens and the plants in them into separate pigeonholes. Until recently, vegetable gardens have been kept separate from ornamental plantings. Fruit trees and herbs have occasionally been included in garden designs that rely almost exclusively on inedible ornamental plants; vege-

A rectangular patch of vegetables in midlawn.

In this garden, vegetables were planted in a flower border. The combination consists of marigolds, peppers, and flower kale. (Anita and Jim Green's garden, Corvallis, Oregon.)

tables have been grown in rows, often in rectangular patches in the middle of lawns. Some landscaping books have even suggested that they be included in service areas with garbage and laundry lines and separated from ornamental plantings by fences.

Recently, however, there have been some changes. There are a lot more vegetable gardens now than there were ten years ago. According to recent figures, thirty three million (out of seventy nine million) households now have vegetable gardens. Some gardens are in backyards; others are in sidewalk strips and front yards, where grass has given way to corn and tomatoes. Some are in planters, on patios, roofs, and window ledges. Two million gardens are plots within community gardens.

Many of these gardens are small, and tight space encourages creativity. Some gardeners have begun planting flowers in vegetable gardens and vegetables and herbs in flower borders. When we stop thinking of food plants as an isolated category, we move back toward age-old traditions. All over the world, most gardeners have grown only the plants they could use, and they have grown them all together—herbs and flowers for seasoning, medicine, and scent; vegetables, fruits, and nuts for food. This custom persists in many countries today, and it can transform American gardens.

Our gardens can be whole! We can live surrounded by plants that are beautiful both aesthetically, and in the oldest, deepest way known to us—because we need them.

# The Beautiful Food Garden

# Chapter 1

# Food Plants As Part of the Landscape

Imagine yourself sitting under a trellis of grape vines, in a patio bordered by planters full of basil, lettuce, nasturtiums, marigolds, and snapdragons. The fragrance of the plants surrounds you; if you reach out, you can pick and eat some of them. Nearby is a pool with water chestnut and arrowhead leaves rising above the water and water lilies floating on its surface. Across a spicy-scented carpet of creeping herbs, ripening fruits hang from trees and shrubs. And herbs, vegetables, and flowers grow in sunny spots and borders among the trees.

This book is a practical guide for growing such a garden—a garden that is lovely to look at and that also provides you with food, a garden where inedible ornamentals and attractive food plants grow together. You can grow ornamentals and attractive food plants together any place you have some sun and soil (or even a hydroponic setup) in which to garden.

## CONTAINER GARDENING

If all your plants are in containers—on a patio, window ledge, balcony, or roof—you have some special limitations and advantages. One limitation is that you need to water and fertilize plants more frequently than if you were growing them in open soil. You also need to protect perennials more carefully during cold winters. If you want to grow food plants that become large, such as corn, or deep-rooted, such as carrots, you may need to use special midget varieties. You may want to avoid food plants that are not especially attractive, such as spinach and turnips, since in container gardens there are few out-of-the-way corners to hide anything and you want every inch of your garden to be attractive as well as productive.

You also have some real advantages because you are growing plants in movable containers. For example, you can sow seeds indoors in permanent containers and move seedlings outdoors without disturbing their roots by transplanting. You can drag containers to the background when plants are young and insignificant or old and worn-looking. If you are growing a different type of plant in each container, you can move the containers around until you have plant combinations you like. Then, at the end of the growing season, you can easily bring frost-tender plants indoors for winter use and complete protection from cold. You can move perennials that need slight winter protection close to house walls for warmth, and you can wrap and mulch the containers.

In places where winters are wet and soils get soggy, container gardens with a good soil mix drain well after fall rains and dry out faster for early planting the next spring.

## SMALL GARDENS

When space is tight, mixing attractive food plants and ornamentals can help you get much more beauty and enjoyment from the space that you have.

You can plant dwarf fruit trees, compact fruit bushes and vegetable varieties, and climbing food plants that use vertical space. (See figure 1-1). If you have time, you can take advantage of labor-intensive methods of gardening that might wear you out if you tried them in a larger garden. By sowing early indoors or in cold frames, and by succession planting, (planting one crop after another in the same place during the same growing season), you can expand the growing season to get a lot more food and flowers. You can sow thickly and harvest plant thinnings. You can greatly improve the soil by adding lavish amounts of well-decomposed organic matter. Also, because you may have more time to take care of them, you can plant beautiful flowers that require frequent trimming.

## CITY GARDENS



If your garden is a small dooryard in a big city, you may want to cram it with all the food plants you can fit in. One thing you probably do *not* have to worry about is whether or not your neighbors will approve of

1-1. Runner beans use vertical space and form the walls of an outdoor room. (Maryanne Caruthers-Akin's garden, Portland, Oregon.)

your choice of plant materials. If you put in carrots instead of petunias, for example, why should they care? They will be glad that you planted something green to look at, and chances are they will either ask for a taste of your carrots, plant some themselves, or ask you where you got those "neat ferns."

Your garden may be shaded by tall trees and tall buildings. This limits, but does not destroy, your choice of food plants. Mint, lemon balm, and day lilies do well in shady conditions, and many leaf and root crops grow with adequate, if not spectacular, vigor if they are shaded part of the day. You may find that, although ground-level gardening conditions are shady, upper parts of your building are sunny. Containers and climbing plants on window ledges, balconies, and roofs can take advantage of this light.

If vandalism is a problem, avoid food plants with inviting fruits. Tomatoes, watermelons, and pumpkins seem to be the most attractive to vandals.

Testing recently began in some cities to determine lead content in city-grown vegetables. Lead in vegetables can come from the soil, especially on sites where buildings painted with lead-based paint have been demolished; it can come from the air in areas where automobile exhaust is heavy. Lead content is highest in leaf vegetables, medium in root crops, and lowest in fruits (tomatoes, beans, corn, squash, etc.) Soils and vegetables from some suburban sites do not show dangerous levels of lead, but the amount of lead in a medium-sized serving of leaf vegetables from some city gardens approaches the maximum safe level for children, who absorb four times as much lead as adults and suffer brain damage and other health problems as a result of eating too much of it. The gravest danger to small children comes from playing in the soil of lead-contaminated gardens.

Washing vegetables, and especially soaking them in vinegar and water, can remove a good deal of surface lead. Peeling removes surface lead from root vegetables. Don't mulch your garden with leaves or grass that have been lying in the street (or that grew near heavily traveled streets). Plant barrier hedges between the garden and nearby heavily traveled streets. All food gardens should be sited as far as possible from trees, and away from lots where buildings with lead-based paint were demolished. For further information read the Boston Urban Gardeners' *Lead Booklet.* (See Bibliography). If you are worried about lead content in your soil, stick to ornamentals and fruit-bearing crops until you can get your soil tested. If you think your children may have eaten lead, ask your doctor to have their hair or blood tested for lead content. It costs about thirty dollars (as of 1981).

Cadmium is the other heavy metal that is dangerous to urban vegetable gardeners. Cadmium causes kidney damage and cancer. One important source of it is local industries that use or discharge cadmium through their processing, thus polluting the soil and water. Another important source is sewage sludge that may be contaminated with industrial cadmium. Unless the sludge you use has been tested for heavy metals and you are sure it is "clean," don't use it for fertilizer on your garden. (Lead can come from these sources also.)

If you want to have your soil tested for lead and/or cadmium, ask the Cooperative Extension Service and Environmental Protection Agency if they can do it. They may say no, but ask. In Massachusetts there is a free soil test (and blood test) for lead contamination. For thirty-two dollars (as of 1981) the Soil and Health Society, 33 East Minor Street, Emmaus, PA. 18049, will test soil for lead and cadmium (plus soil nutrients and pH).

## COMMUNITY GARDENS

Many community gardens are a focus for neighborhood life. Plays, fairs, parties, story readings, and casual socializing take place there, as well as gardening, and an edible landscape can provide a wonderful environment for these activities. Community gardens are often divided up into rectangular plots for each individual or family. Even within a rectangular format, the plants in your plot will look more attractive if they are grown in beds, rather than rows. Vegetables that people use a great deal can be grown in individual plots; fruit trees, herbs, ornamentals, and perennial and low-volume vegetables can be grown in beds that everybody shares.

1-2. Left: plum tree along a fence. (Photo by George Gessert, Jr.) 1-3. Right: black walnut in a street scene, Eugene, Oregon.

A community garden can be used as a matrix from which whole neighborhoods can become beautiful and productive. It can include cold frames and a greenhouse for starting seedlings, as well as a nursery for perennials and fruit and nut trees. The trees can be planted along lightly-traveled streets for neighborhood sharing (if city street tree regulations allow this), or along property boundaries for sharing among families. (See figure 1-2.)

## FRUIT AND NUT TREES AS STREET TREES

City regulations in many areas do not allow fruit and nut-producing trees along the street, because fallen fruits, nuts, and nut husks can clog drains and create a hazardous ground surface. In many cities, anything you plant along the street legally belongs to the city, so, if officials do not like what you plant, it is within their rights to chop it down. (In countries that have less food to spare, these rules of ours would seem very peculiar.) If and when the American food and food transportation situations change as a result of fuel shortages, expense, and other factors, we may regret not having had the foresight to plant these trees on our streets.

I feel that fruit and nut-producing plants should be planted on the streets now, especially in densely populated cities where people take great satisfaction in growing their own food and becoming self-sufficient in whatever ways they can, and where there are often plenty of poor people who really need the food source. Nut trees are especially important. (See figure 1-3.) Although nut crops are sometimes uneven, they are easy to store and are loaded with protein, calories, vitamins, and minerals; they can be a staple article of people's diets.

If local rules forbid fruit and nut-producing trees on the street, try to get the rules changed or to negotiate. City officials may be impressed if you organize a formal system of tree maintenance and harvest, such as a children's tree corps, or a network of individuals willing to contract themselves as volunteer caretakers for certain trees.

## SUBURBAN GARDENING

If you are gardening in the suburbs and need more space, there may be more available than you think. The sunniest and least used part of many suburban properties is the front lawn. You can let the grass grow a bit and add fruit trees and meadow wildflowers for an old-fashioned orchard. You can plow the lawn under and grow a food garden in the summer; in winter, you can grow cold-hardy vegetables and herbs, or make a lawn again by sowing a winter-hardy cover crop. If you are devoted to your lawn, you can keep it and add mowable, ground-cover herbs—creeping yarrow, creeping thyme, and Roman chamomile.

Occasionally, neighbors object to front-yard food gardens. You can ignore their objections and wait for them to get used to the idea, or woo them with fresh vegetables.

Border strips between the sidewalk and street are often occupied only by weeds or bark mulch. If there is little traffic going by (and little lead being discharged), these strips are good places for fruit trees and vining vegetables. If dogs are a problem, plant crops that are hard-shelled or borne well above the ground.

## LARGE GARDENS

If you have a large garden, you already know it must be carefully planned so that maintenance is within human capacity. You may want to concentrate on perennials and long-term annuals so that you will not need to do much succession planting. You may avoid thick plantings that need to be promptly thinned and prolific short-lived flowers that need frequent trimming.

You can grow rambling food plants, such as squash, and you may even have room for something I consider a real luxury item, space-wise—ugly food plants. Some food plants and plant varieties are not attractive but produce food that is better or substantially different from comparable attractive food plants. Dry beans and mildew-prone squash varieties, for example, are good choices for a row garden in an inconspicuous place. You may also have room to grow plants with limited food uses, such as chamomile for tea. You may be able to experiment with ornamentals that have exotic food uses, such as redbud, a beautiful tree whose juicy flowers can be cooked or added to salads. (At the end of Chapter 3 are lists of Less Attractive Vegetables and Herbs, Plants with Limited Food Uses, and Possibilities to Explore, which includes ornamentals that have exotic food uses.)

## CHOOSING PLANTS FOR THE BEAUTIFUL FOOD GARDEN

You may want a garden that is entirely edible, where every plant is both attractive and productive, or a garden that combines food plants and standard ornamentals. Many "edible landscapes" are better described as partly-edible landscapes; they have plenty of food plants, some inedible flowers to provide extra color, and inedible woody perennials to help give the garden strong year-round structure.

Some purely ornamental plants provide effects few food plants can equal. Broadleaf evergreen shrubs, for example, are inedible, and some are even poisonous, but they are important if a garden is to look good in winter. Most spring-flowering bulbs are inedible. Certain inedible ornamentals have important food-related uses. They are used as companion plants or as a food source for wildlife. We all have plants we especially like; we feel our gardens would be incomplete without them. Some of them are plants that seem very beautiful or interesting to us, and some are full of personal associations. Probably not all of them are food plants.

When you choose plants for your garden, think about what type of plant is needed for a specific environmental niche and purpose in the garden, and see if there is a food plant that will fit in that situation. In some cases, there is no food plant that will work, and an inedible ornamental fits perfectly. But in many other cases, a food plant is just as satisfactory as the purely ornamental alternatives, and you will have the added pleasure of being able to use it for food.

When you choose an attractive food plant for your garden, consider whether you like the food plant enough to get around to eating it. What will happen if you don't eat it? (Maybe you decide you don't like it, or you forget it is there.) If you ignore a plum tree, there will be a foul-smelling mess of rotten fruit, but if you do not pick the currants on a currant bush, or the leaves on kale, the plant will still be an attractive part of the garden.

When considering a particular kind of attractive food plant, you must decide how many plants you want. If growing space is ample, you may deliberately plant more of something than you are going to eat,

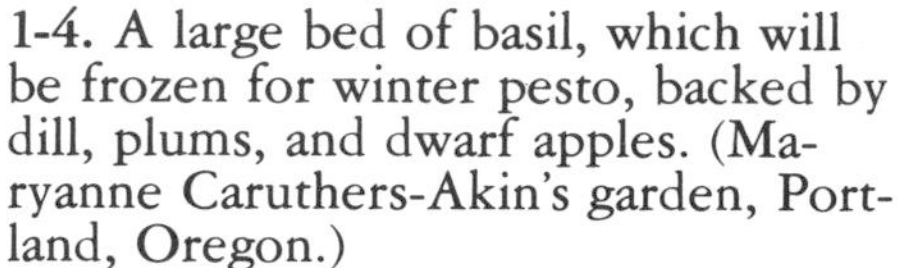

1-4. A large bed of basil, which will be frozen for winter pesto, backed by dill, plums, and dwarf apples. (Maryanne Caruthers-Akin's garden, Portland, Oregon.)

because you like the way it looks. For example, you may need a ground cover for an eroded bank. Instead of ivy you decide to plant thyme, the plant you feel is best both for the visual purpose you have in mind and for the environmental conditions available in that place. It is unlikely that you will eat all the thyme growing on the bank, but you will certainly eat more thyme than you would ivy!

If you become involved in edible landscaping and have a large garden, you may come to have so much productive garden space that you can feel flexible about whether or not you use all the food you grow. You can share your windfalls with the neighbors or preserve some for winter. (See figure 1-4). Some years you may be very efficient and use everything; other years, you won't. But most importantly, you will have an attractive garden, and the food will be there when you want it.

Note: There are some plants in the encyclopedia that you may not have eaten before; eat a small quantity of them before you eat a large quantity to make sure you do not have an adverse reaction to them, especially if you are allergic to any other foods.

# Chapter 2

# Designing the Beautiful Food Garden

Most American yards have the same components: decorative front-yard plantings, driveway, patio, lawn, shade trees, service area, and sometimes a children's play area and a vegetable garden. This sameness would be all right if we all had the same activities, but obviously we do not.

Many people embellish the standard format by driving to the local nursery, choosing some nice-looking trees and bushes, and planting them around the house. Although this spontaneous approach is popular, it can lead to mistakes that are annoying, expensive, and hard to correct. For example, you are likely to pay higher utility bills if you do not consider the microclimate of your site before you put in permanent plantings. Plants in the right places can help your house stay cool in summer and warm in winter.

A careful design can save you trouble and expense. You can try ideas and make mistakes on paper, where they are easily remedied. You can plan space that fits your individual needs and those of your family. You do not have to be an expert or have a good "artistic sense"; the main requirements are time, interest, and patience. A careful design can reward you with a garden that is beautiful to look at and comfortable to live in. As you work on the design steps in this chapter, you may gain a deeper understanding of your and your family's lifestyles and a satisfying intimacy with your garden site.

If your garden is already working well and looking good and you are interested mainly in the final details of fitting food plants into an existing design, you may want to skip to the end of this chapter. (See Step 3—"Choose Specific Textures and Plants"—on page 32.)

If you are interested in making an overall plan for your yard, the design process presented in this chapter may be useful to you. It will help you analyze your family's activities and the garden site separately. Then it will lead you through a number of steps before the two are combined in a final plan. This process can be used to design space for bare lots, for established gardens that need improvement, for big country gardens, city patios, or community gardens. I enjoy it. It is a long process and often I become frustrated in the middle. But, when I finish a design, I feel satisfied that it comes at least close to fitting the site and meeting people's needs, even if it's not perfect. (The design process makes one humble!) The process is used by many landscape architects; I learned it from Steve Lindsey and Chuck DeDeurwaerder, landscape architecture teachers at Oregon State University.

The following materials are needed:

- a 50-foot (15-m) measuring tape
- an architect's ruler or a regular ruler
- a long roll of 18- or 24-inch (45- or 60-cm) architect's tracing paper
- drafting tape
- a wide table on which to work

## ACTIVITY ANALYSIS

Think about the way you will use your site. For now, deliberately disregard the actual scale and details of your site (this is hard if you are living there!) and concentrate on the things that you will be doing there and that you would like to do there.

There are good reasons to disregard your site, for the time being. If you are constantly aware of its limitations, you cannot give full scope to imagining the things you would like to do there. (Obviously, you are bound to be somewhat aware of these limitations. For example, if you love to play tennis and have only a small patio, you can't play tennis on your site and there's no point making a design that incorporates tennis courts. But for less extreme discrepancies, you are more likely to imagine all the wonderful things you'd like to do on your site and find ways to do them later in the design process if you try to think only of your activities in the beginning.) Disregarding the site also prevents you from making strong, immediate commitments to certain plants and certain aspects of the site. Such commitments can get in your way as you move through the design process.

### Step 1. Make an Activities List

List in any order every conceivable activity that might occur on the site. Be playful, be flexible, take risks. You need input from everyone who will be using the space regularly. Consider outdoor space alone, or, if you wish, indoor space as well. (Your house is probably already built, but you still have some control over which activities take place where.)

### Step 2. Make a Qualities List

List the qualities important to you in your environment, for example, privacy, energy conservation, attracting wildlife, safety for children, food production, a feeling of peace. These qualities will not appear as geographical entities on your plan, but you can refer to them throughout the design process.

### Step 3. Make Activity Groups or Balloons

Take the activity list from Step 1 and group together activities that are related to each other. Write them down on heavy paper and cut out each group as a separate balloon shape. Give each group a generalized name and an approximate size relative to other groups. (See the example in figure 2-1. The people who are designing this space live in a city with a wet, mild climate. There are two adults and two small children.)

Begin thinking now about where and how you want to grow food plants. You may plan to combine food plants with purely ornamental plantings, but also allocate special areas where food plants will be prominent and have ideal growing conditions, such as "frequently used plants" in figure 2-1.

### Step 4. Make a Balloon Arrangement

Once you have cut out activity group balloons, begin moving them around in different combinations to see which balloons need to be near other balloons. One concession to reality (to your site as it exists) does need to be made here. Make a balloon for your house (and another one for the street) and place these where they actually are in relation to each other. Then the balloons that need to be near the street or near the house (or near specific indoor rooms that you might want to note within the house balloon) can be placed close by. An outdoor eating area, for example, needs to be near the kitchen.

The balloon "frequently used plants" (See figure 2-2) needs to be near "tools" and "quiet space" (so you can enjoy the plants up close

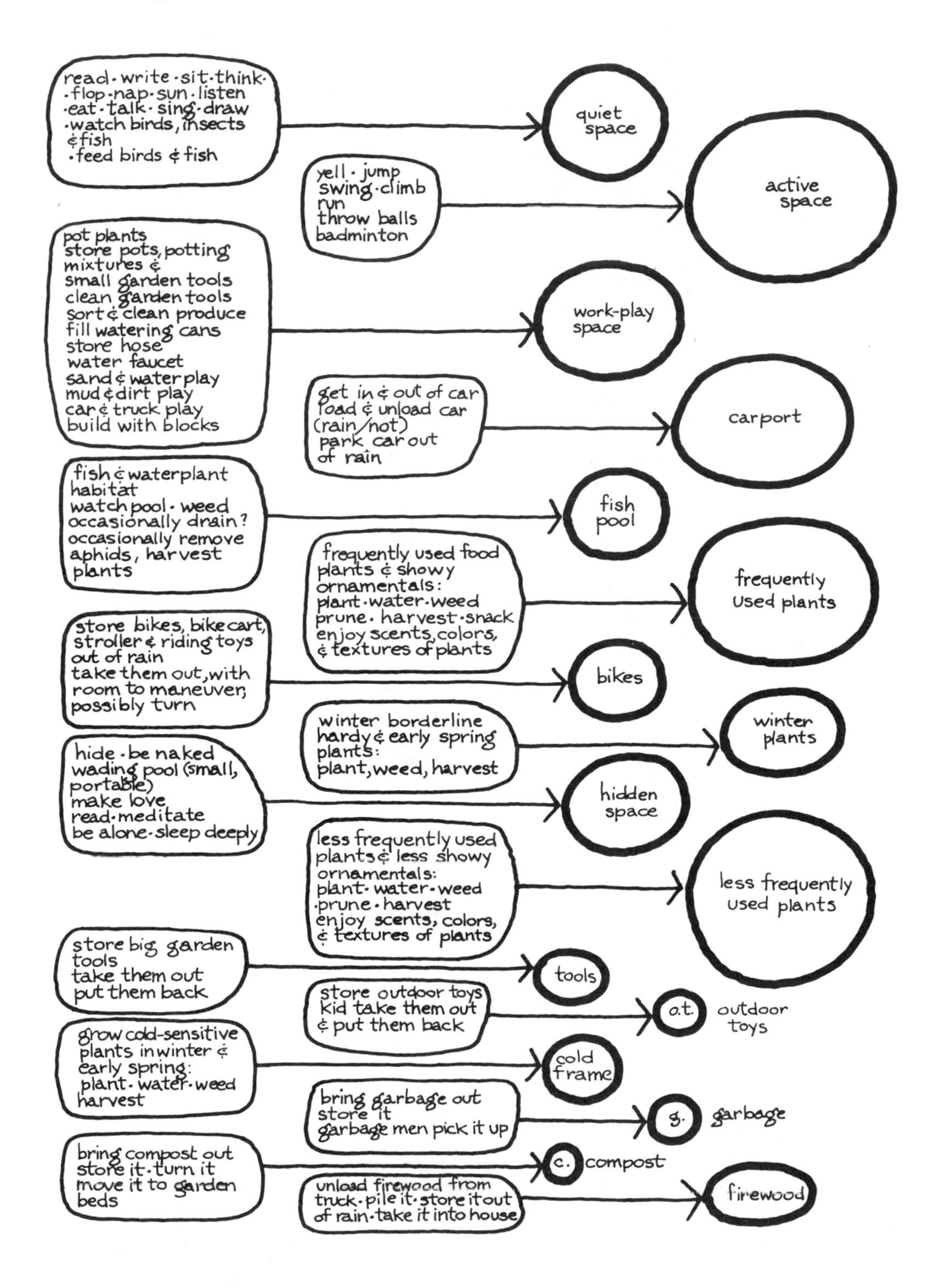


2-1. Activity analysis: Step 3. Grouped activity lists (left) become balloons of approximate relative sizes (right).

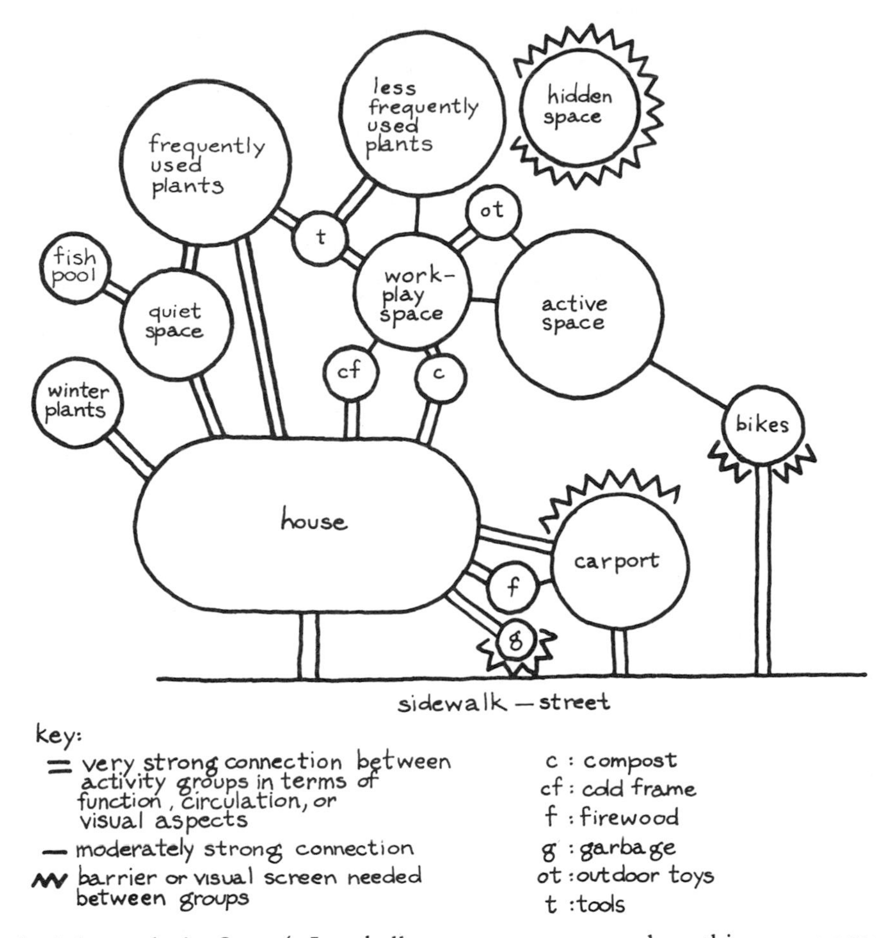


2-2. Activity analysis: Step 4. In a balloon arrangement such as this one, a generalized house and sidewalk-street are included as reference points, since some activities need to be near them.

while relaxing). Frequently used plants do not need to be right next to the house unless your site is very large.

When you find an over-all arrangement with good possibilities, lay tracing paper over the balloons and trace it. Once you have recorded the arrangement, you may notice problems with it. Scribble suggestions on the margins of the paper, keep it to refer back to, and start moving the balloons around again to see how you can improve on it. When you do, trace it.

There is no ideal solution, but the more ideas you try out, the more likely you are to reach an arrangement that satisfies you. Then move on to the next step. Take your time; if you go slowly, you are more apt to be satisfied with your design.

## Step 5. Make Circulation Patterns



Circulation routes are spaces that you go through as you move from one activity group to another. Tape more tracing paper on top of your

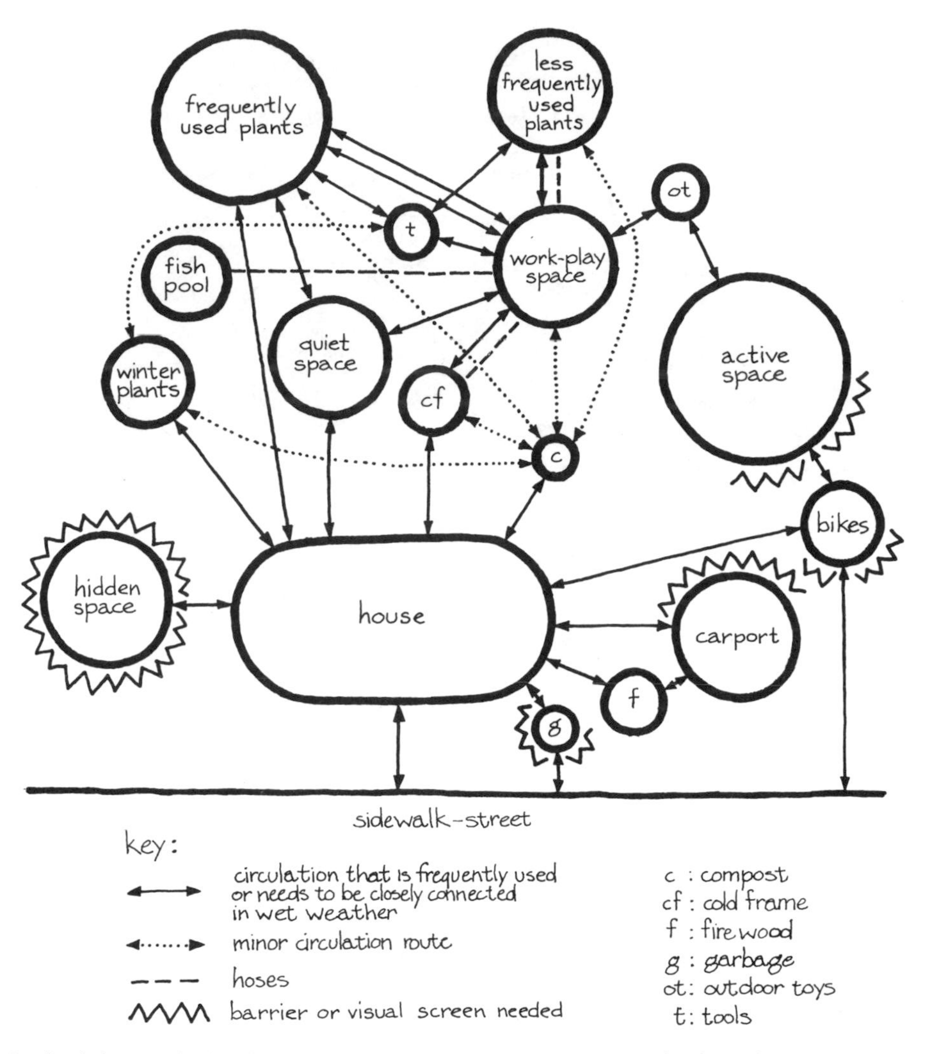


2-3. Activity analysis: Step 5. Circulation patterns. (Final drafting by George Gessert, Jr.)

Step 4 plans, and make any changes that are necessary for good circulation. (See figure 2-3.) Which routes need to be more direct? Which can meander? Which are main routes, and which are used only occasionally? You can use solid lines between balloons for important circulation patterns and dotted lines for minor ones.

The balloon "frequently used plants" has strong circulation connections to the house, especially the kitchen, as well as to "tools," "quiet space," and "work-play space," with its water source. If plants will be frequently used in winter as well as in summer, circulation on dry paths becomes important. A food plant area should *not* be easy to reach from an active play area.



## Step 6. Estimate Size of Activity Groups

Figure out how much space your various activity groups require—both horizontally and vertically. A good scale for designing most residential and small-scale sites is ¼ inch to every foot (an approximate metric equivalent of this scale is 4 cm to 1 m).

Landscaping books sometimes include tables that estimate how much space to allot for various outdoor activities, but you can reach more satisfying solutions if you make your own estimates, using a tape measure, your imagination, and your body. Move around acting out how much space is required when you get up from a table during dinner, for example, or how much space is required for pushing a wheelbarrow around raised beds. Measure pieces of furniture or equipment that are likely to be involved and estimate ideal measurements for things you do not yet have. Provide space for circulation as well as activity groups; wide paths lend strength and convenience to major circulation patterns.

How much space you allocate specifically for food plants depends on three factors: (a) how much food you like to grow; (b) how intensively you use space (closely planted areas that are in use throughout the four seasons produce more food per unit of space than do sparsely planted areas used during the summer only); and (c) how much you integrate food plants and make them work in the landscape (for example, if you intend to plant dwarf fruit trees as a hedge instead of privet, you will have less need for separate "orchard" space).

### Step 7. Design Shapes of Activity Groups

Now design shapes for each activity group. Some space shapes are suggested by the activities themselves. For example, people will eat in "quiet space," and you already have a hexagonal outdoor table and six chairs, so you may want to make the "quiet space" hexagonal. Some space shapes may be suggested by their relationship to adjacent spaces. Others are more a matter of choice. Experiment until you find shapes that suit the activities that happen in them and also work well as parts of the whole. Make a pattern of shapes that pleases you. (See figure 2-4.) (If you feel uncomfortable in this step because things are too open-ended, you can try using one of the "shape-themes"—rectilinear, angular, or curvilinear—described on page 30.)

Once you have shaped scaled activity groups in a pattern that satisfies you, you are through with activity analysis which is in some ways the toughest part of the design process.

## SITE ANALYSIS

This part of the design process is sensory and immediate. It means getting to know a place thoroughly—its practical elements and its aesthetic elements, the spirit of the place.

John Ormsbee Simonds, author of *Landscape Architecture: The*

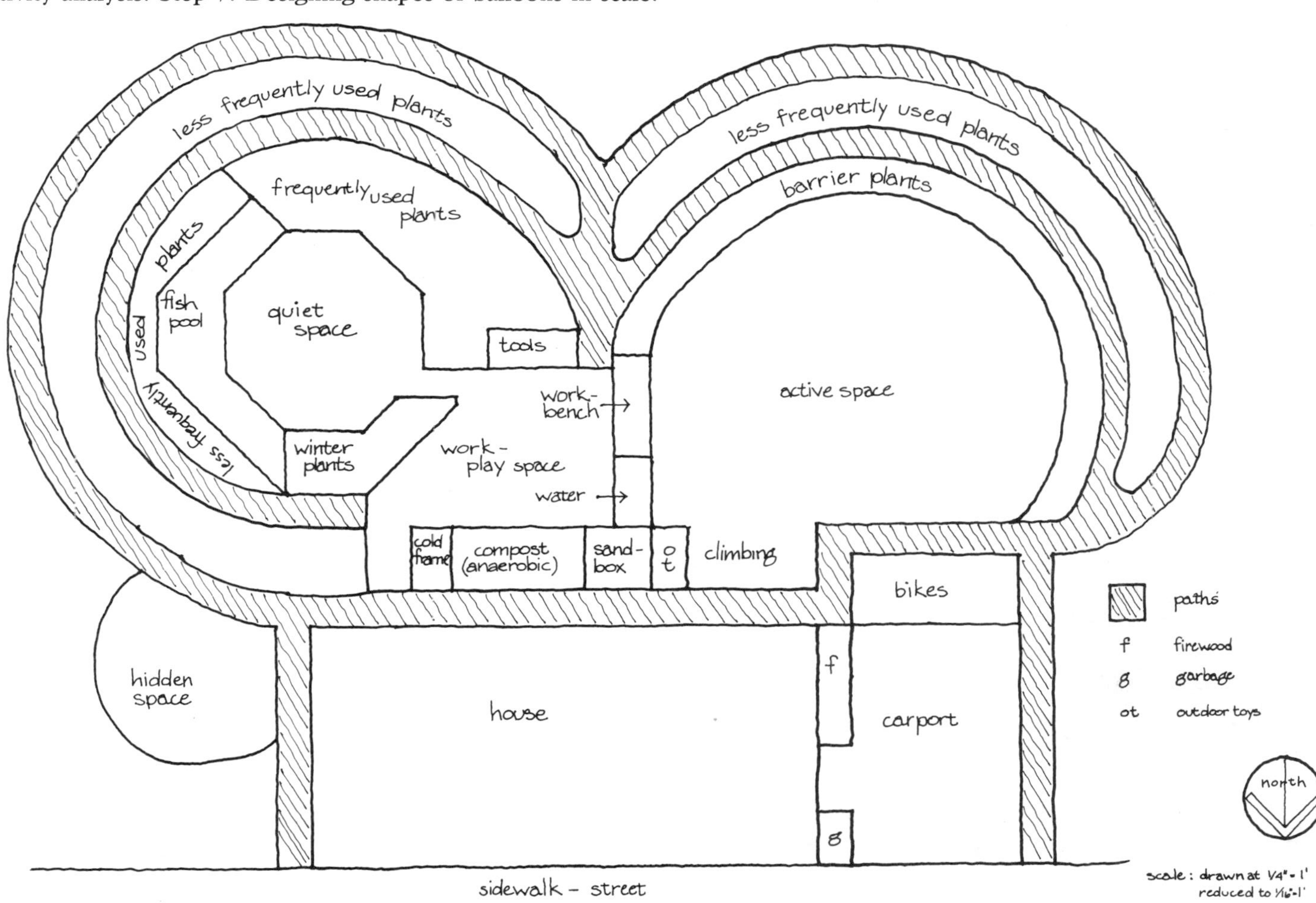


2-4. Activity analysis: Step 7. Designing shapes of balloons in scale.

*Shaping of Man's Natural Environment* (New York: McGraw-Hill Book Company, 1961), once asked a Japanese architect how he made his designs so sympathetic with their sites. The architect answered:

> If designing, say, a residence, I go each day to the piece of land on which it is to be constructed. Sometimes for long hours with a mat and tea. Sometimes in the quiet of evening when the shadows are long. Sometimes in the busy part of the day when the streets are abustle and the sun is clear and bright. Sometimes in the snow and even in the rain, for much can be learned of a piece of ground by watching the rain play across it and the run-off take its course in rivulets along the natural drainage ways.
>
> I go to the land and stay, until I have come to know it. I learn to know its bad features—the jangling friction of a passing street, the awkward angles of a wind-blown oak, an unpleasant sector of the mountain view, the lack of moisture in the soil, the nearness of the neighbor's house to an angle of the property.
>
> I learn to know its good features—a glorious clump of maple trees, a broad ledge perching high in space above a gushing waterfall, which spills into the deep ravine below. I come to know the cool and pleasant summer airs that rise from the falls and move across an open draw of the land. I sense, perhaps, the deliciously pungent fragrance of the deeply layered cedar fronds as the warm sun plays across them in the morning. This patch I know must be left undisturbed.
>
> I know where the sun will appear in the early morning when its warmth will be most welcome. I have learned which areas will be struck by its harshly blinding light as it burns hot and penetrating in the late afternoon, and from which spots the sunset seems to glow the richest in the dusky peace of late evening. I have marvelled at the changing, dappled light and soft fresh colors of the bamboo thicket, and watched for hours the scarlet-crested warblers who nest and feed there.
>
> I come to sense with great pleasure the subtle relationship of a jutting granite boulder to the jutting granite profile of the mountainside across the way. Little things, one may think, but they tell one, "Here is the essence of this fragment of land; here is its very spirit. Preserve this spirit and it will pervade your gardens, your homes, and your very life."

2-5. "Here is the essence of this fragment of land."—John Ormsbee Simonds

2-6. The church tower across the street: St. Peter–St. Paul, Our Lady of Pilar, Court Street, Brooklyn, New York.

I disagree with the Japanese architect in one respect—his feelings about cities. What he calls "the jangling friction of a passing street" is not necessarily a bad feature, especially if the passing is mostly friendly and pedestrian. If you live in a city you like, you may find that you want some parts of your site to *welcome* the city's nearness (while you will want other parts to be very private). You may also find that the spirit of your site has less to do with "a glorious clump of maple trees" than it has with the old brick wall of the house next door to you, and that your favorite view is not "the jutting granite profile of the mountainside," but the old church tower across the street. (See figure 2-6.)

The city itself and its people can become the spirit of the landscape, especially where cities are one or two hundred years old. Many cities have souls as strong and beautiful as their country counterparts. Getting in touch with the soul of a city (or a neighborhood) can transform your design.

## Step 1. Be Receptive to the Site

Wherever you live, absorb as much as you can about the site and write down anything you want to remember. Do not yet consider what you will *do* with the site; just think about it as it is—its man-made and natural features. Even if you have lived for years on the site, you may still learn new things about it through unstructured receptivity.

Return to this step any time you feel it is appropriate during the steps that follow, when you feel out of touch with the site, or when you are bogged down in the design process and need to regain perspective. You may be surprised to find that when you relax and listen, and get in touch with the site, the design solutions inherent in the nature of the land become visible to you.

## Step 2. Make Scale Maps

Working at the same scale as before, make a map of the site complete with all its existing features. Measure the outer boundaries of the site first and work in from there, locating, say, the corner of a building that is 8 feet (2.4 m) south and 10 feet (3 m) east of the northwest corner of a city lot.

Most information you need for site analysis maps can be gained through observation. Man-made features to be measured and mapped are property boundary lines, existing buildings and their features (windows, doors, wall elevations, general room locations inside a building), roads, driveways, paths, steps, curbs, gutters, other paved surfaces, walls, fences, locations of utilities and meters, sewers, water and gas lines, telephone poles and lines, and nearby off-site buildings.

Other features of the site may be natural or may result from a mixture of man-made and natural forces. These features include good and bad soil for growing plants, slopes, gullies, relative land elevations, sun and shadow patterns, prevailing wind patterns for various seasons and times of day, low-lying frost pockets, existing boulders, streams, pools, trees and bushes, flowers and wildlife at various seasons, big trees growing on nearby sites, noises and smells coming from on and off the site, best and worst views, and microclimatic effects (hot and cold, windy and still, dry and wet spots) on the site. Most of this information comes from simple observation. You can get information on prevailing winds from airports and weather bureaus. For sun and shadow patterns, record these patterns at different times of day in the season when you are making your design, and project what the patterns will be at other seasons—when the sun rises and sets in different positions of the sky, when it is higher or lower above the horizon.

Record all concrete physical features of your site on one map (see figure 2-7), and tape on overlays with sun patterns, wind patterns, etc. (Figures 2-8 through 2-12 are examples of such overlays. For easy reading, each refers to the same section of the site, part of the backyard along the south wall of the house.) As you note these aspects of your site, you will have feelings and reactions to what you notice. Write all of them down as part of the mapping process; they are very important.

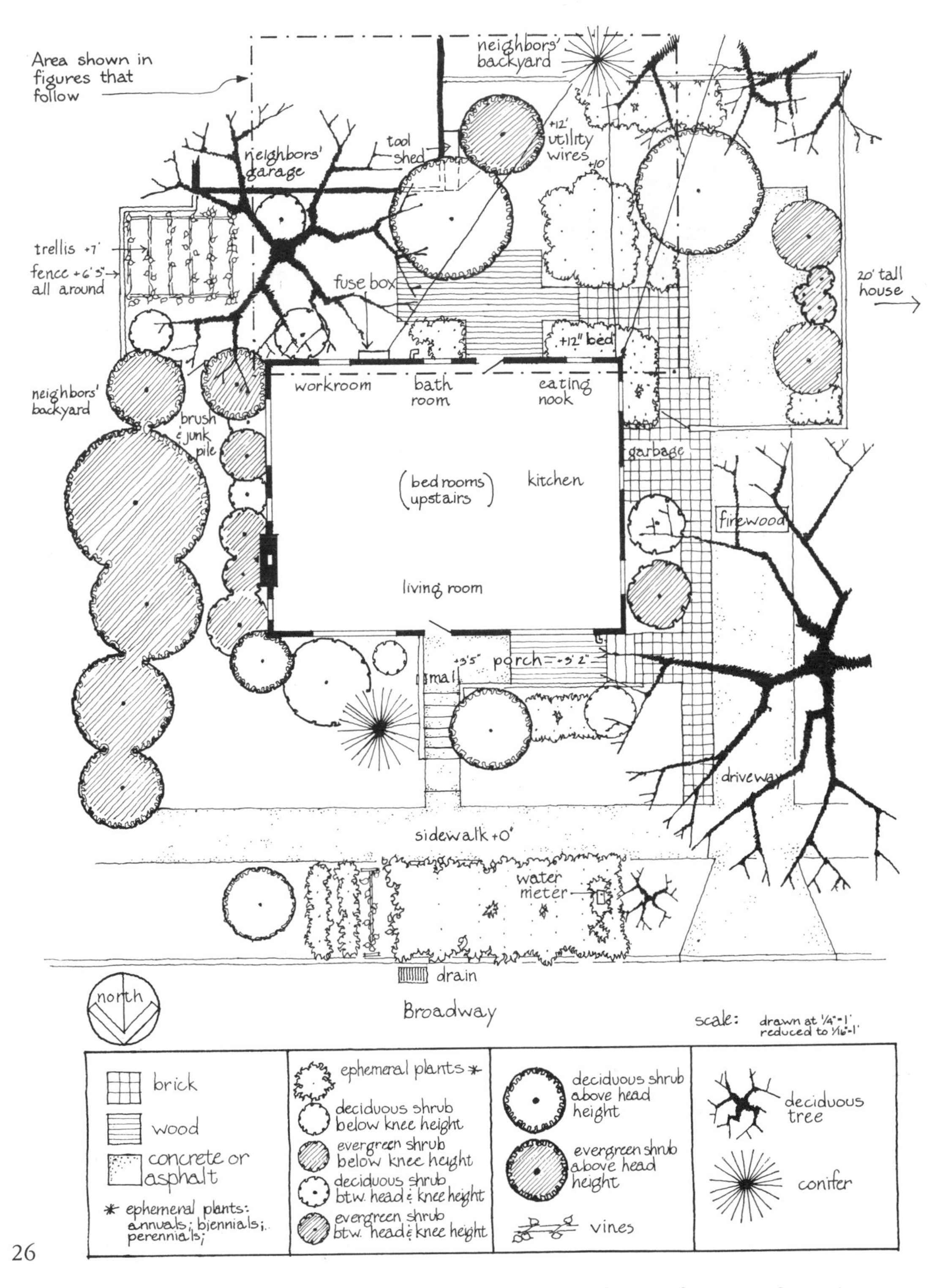




2-7. Site analysis: Step 2. Scale map with concrete physical features of the site.

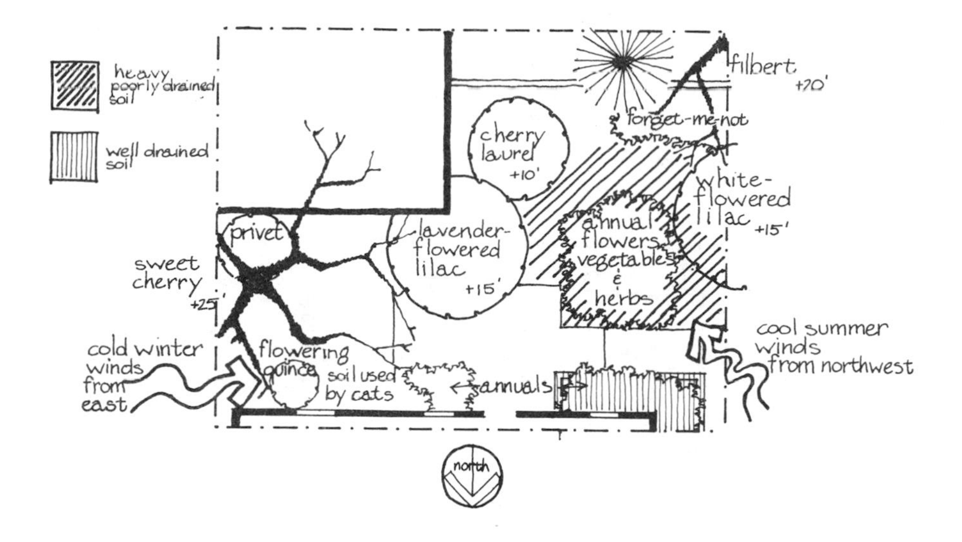


2-8. Site Analysis: Step 2. Area of site in backyard along south wall of house. Overlay 1: specific plants, prevailing winds, and notes about soil.

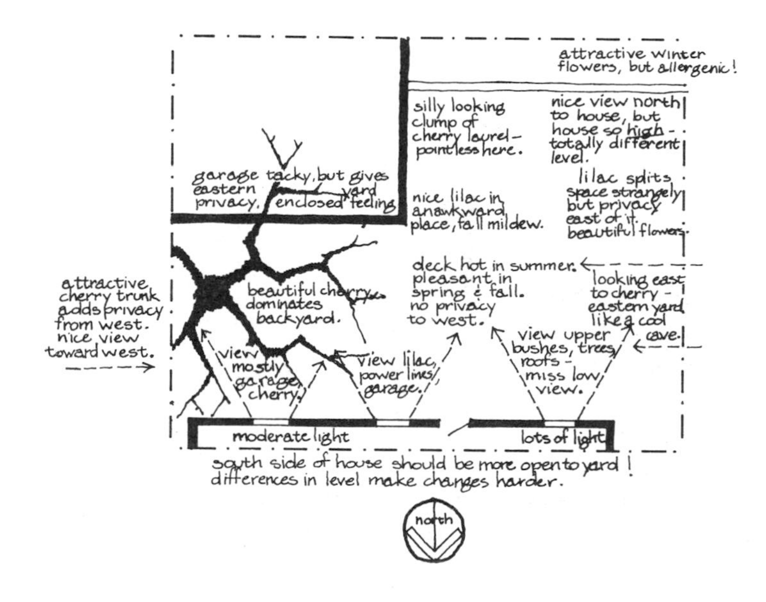




2-9. Site Analysis: Step 2. Same area of site. Overlay 2: views and personal reactions.

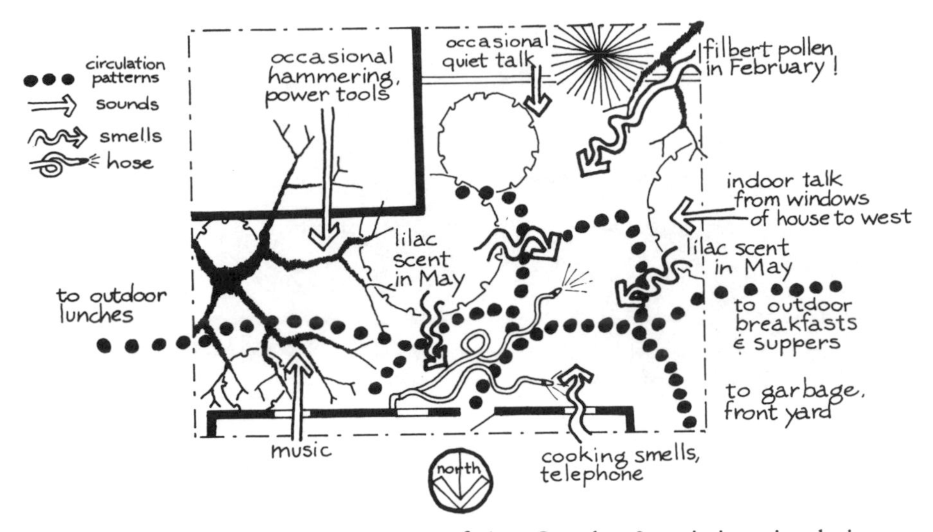


2-10. Site Analysis: Step 2. Same area of site. Overlay 3: existing circulation routes, hose circulation, smells and noises coming from on and off the site.

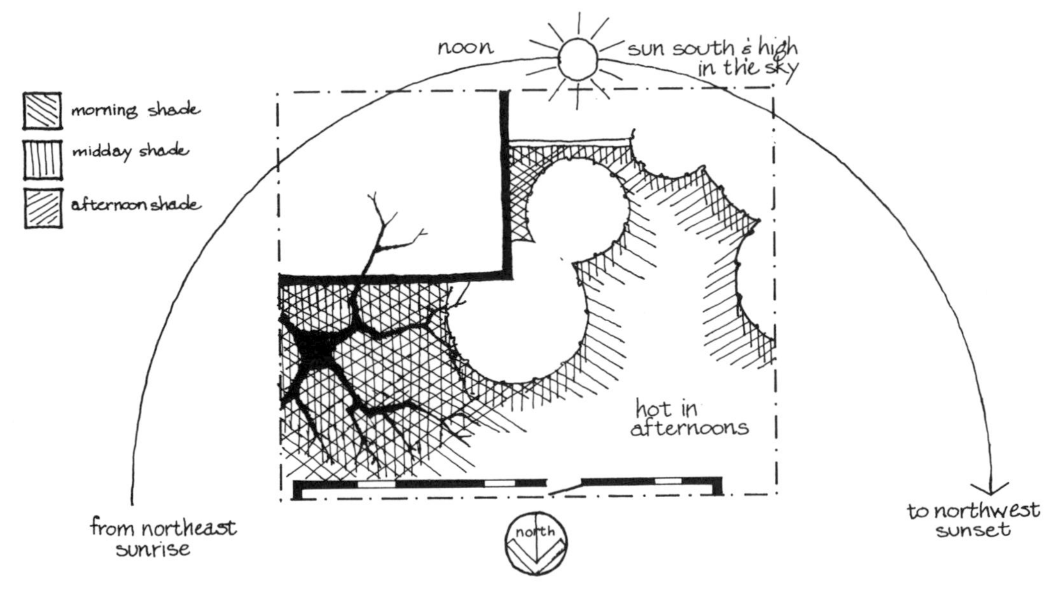

2-11. Site Analysis: Step 2. Same area of site. Overlay 4: sun and shade patterns at the summer solstice.

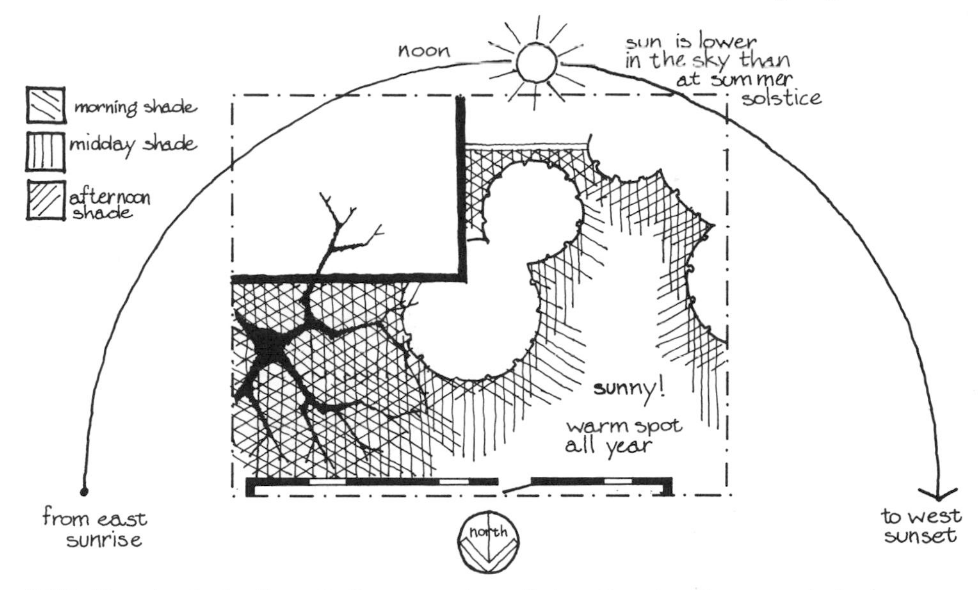


2-12. Site Analysis: Step 2. Same section of site. Overlay 5: sun and shade patterns at each equinox. Midday shadows are longer than those at the summer solstice.

## COMBINING SITE ANALYSIS AND ACTIVITY ANALYSIS

Once you have completely mapped out the various aspects of the site and its uses, you are ready to combine everything you have found out in site analysis with your final design from activity analysis.

### Step 1. Put Everything Together

Place the final design you arrived at from activity analysis on top of the map and overlays from site analysis, and begin to reconcile the two. Shift the activity design around at various angles so it fits in the best possible way on the total site—solid features, wind patterns, and all—and begin drawing a design that incorporates everything.

Roughly indicate where planted areas will be. Plants can modify many features of the site. They can block or filter bad views and draw attention to good ones; control erosion; cut down on noise, smells, air pollution, summer heat and glare, and winter heat loss from buildings; and block or direct winds.

Plants can define or enclose spaces in a design (see figure 2-13), serving as roof, floor, and walls for "outdoor rooms," and they can emphasize circulation patterns. Plants also have many aesthetic qualities that you can use in design—their shapes, colors, textures; the changes they go through; their softening effect on design; the moods they produce; the nonvisual ways they appeal to us (through touch,

2-13. Corn emphasizes the border between a garden and the field that surrounds it. Author's garden, Oregon State University Vegetable Research Farm, Corvallis, Oregon.

taste, smell, the sound of rustling leaves) and most of all their living presence in our world.

As you work on your design, you will have to make plenty of compromises. Sometimes you will have to shift the location of activity groups in relation to each other in order to fit them all on the site. You may occasionally be forced to discard certain activities because they do not fit the scale of the site, but it is more likely that you will find imaginative ways of stretching the potential of the site to *include* your activities.

Undoubtedly you will change sizes and shapes of some activity group areas and make choices about which areas most need, for example, morning sun, or cooling winds in summer. Some of the changes you make will probably be governed by necessity; others will happen spontaneously because you see exciting ways that your activities and the realities of the site intermesh. Try as many different solutions as you can; it takes time and flexibility to reconcile all the different factors with which you are now working. (See figure 2-14.)

As you work on this step, if you feel that you are moving toward a hodgepodge collection of various shapes, you may want to choose a shape *theme* to make your design more coherent. You may want to make all spaces rectilinear (with 90° angles at their corners) or angular (less or more than 90°), or curvilinear. Try several different shape themes in rough form and see which you like best. In general, curvilinear designs and those with a mixture of shapes are hardest to design and construct; rectilinear designs are the easiest. A rectilinear theme is

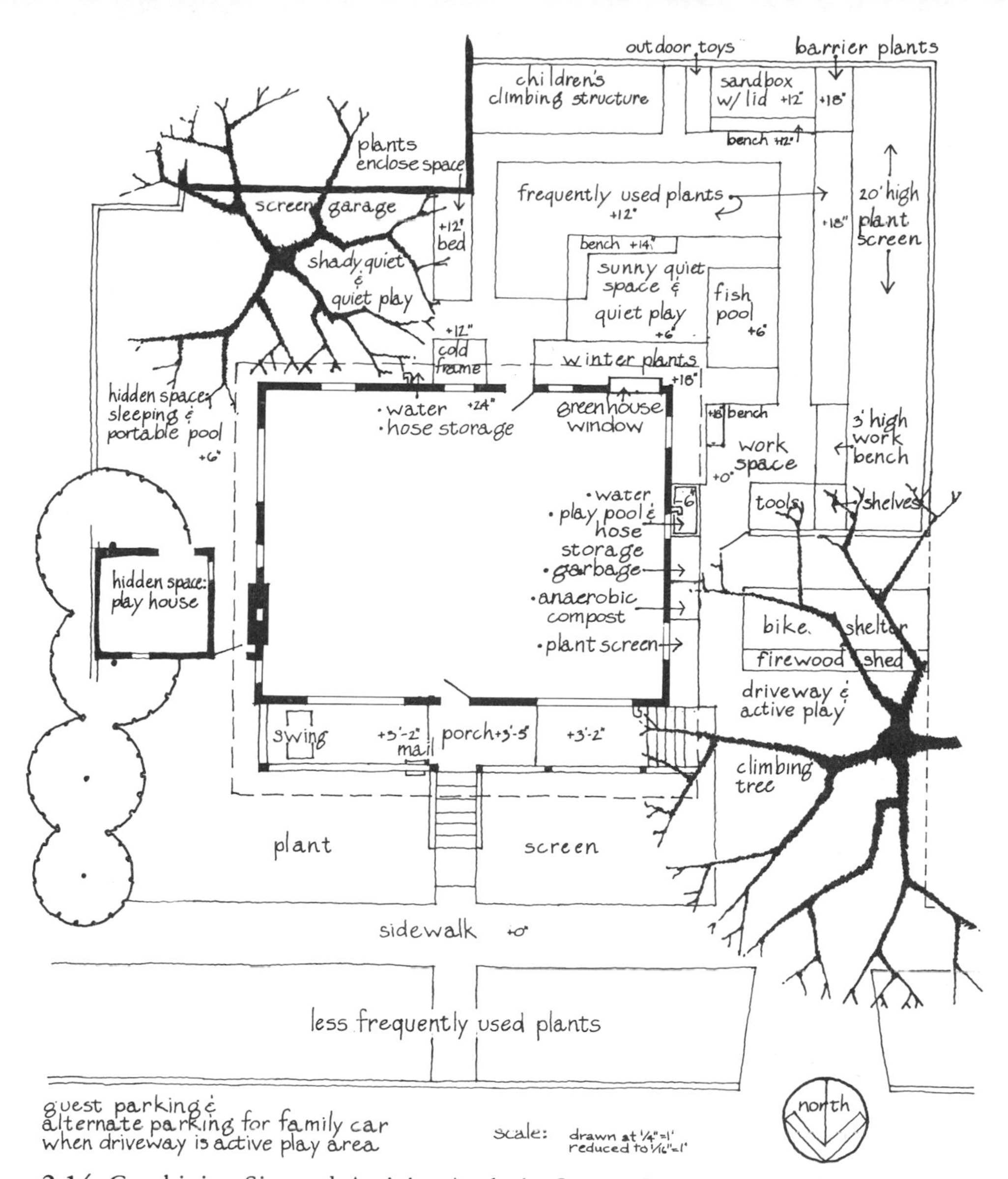


2-14. Combining Site and Activity Analysis: Step 1. Putting everything together. Since Figure 2-4, there have been many changes and compromises. The carport has disappeared and become a driveway, which doubles as active play space for running and ball games. When the driveway is active play space, the car is parked in the street. Climbing is in the big tree by the driveway and also in the south backyard; a swing is on the front porch. To adapt to the shape of the site and to save the sunniest areas for frequently used plants and the fish pool, work space has been separated from play space (except for water play, still included in the work area). The sandbox is along the south fence, and relatively clean, quiet forms of play—car and truck play, building with blocks—have been incorporated in quiet space. Dirt and mud play have been eliminated because there is not room for them in such intensely used space. Quiet space has been split into two areas, sunny and shady. A mail box has been added to the front yard.

There are several outreach solutions to the small size of the site. Infrequently used plants that take a lot of space and/or are grown in large quantities—soybeans, winter squash, etc.—are grown at a nearby community garden. A badminton set is shared with neighbors and set up in their large backyard. Active play space stretches from the driveway west across two neighboring lawns and driveways where the kids on the block all play together. A small amount of anaerobic composting on the site is supplemented by off-site composting done by a local recycling garbage company.

less blocklike than it sounds, because many small irregularities are possible within its right-angle format and plants soften the outlines.

So far, your planning has been done from above, from a bird's-eye view. Now is a good time to add more three-dimensional thinking. You can make rough sketches of your design as seen from the side by a person sitting or standing in it or model some approximate forms in clay or playdough on a board.

A realistic sense of scale also is important. A small site will look fussy if it is divided into too many little spaces. Measure out some elements of your plan on the site itself. Indicate surface outlines and heights by some combination of ropes, chalk outlines, sticks, poles, cardboard, and sheets.

## Step 2. Consider General Textures and Forms

Decide whether ground surfaces of the design will be hard—wood, brick, etc.—or soft—grass and ground cover. Decide whether plants in various places will be deciduous or evergreen. (See figure 2-15.) Think of plants in groups rather than as individuals. Gardens with one each of many different kinds of plants are often spotty-looking.

You will find the plants in your design usually fit in a certain place for a specific reason: a broad-canopied tree shades a patio area; evergreen shrubs block a bad view or provide a barrier between a lily pool and an area where children play.

In many parts of the garden a balance of evergreen and deciduous trees and shrubs will be most effective, with plenty of herbaceous plants added in. Broadleaf evergreens and cone-bearing, needle-leaf evergreens give strong year-round bulk to a garden design, but look overpowering in large masses. Deciduous trees and shrubs contribute color, change, and interesting winter branching patterns, but all-deciduous plantings look bare in winter. Herbaceous plants (those without much woody tissue)—annuals, biennials, and perennials—often die back to the ground during one or more seasons and need to be backed up by woody plants.

There are ways of making herbaceous plantings look more substantial year-round—winter-hardy and early spring-flowering herbaceous plants extend seasons of growth and interest (see figure 2-16); hardy cover crops fill space and enrich soil in off-seasons; brick or wood-sided raised beds are structurally interesting even when bare of plants.



## Step 3. Choose Specific Textures and Plants

At last you can make specific choices! Will an outdoor eating area be paved in brick, wood, stone, or concrete? Where there is room in your

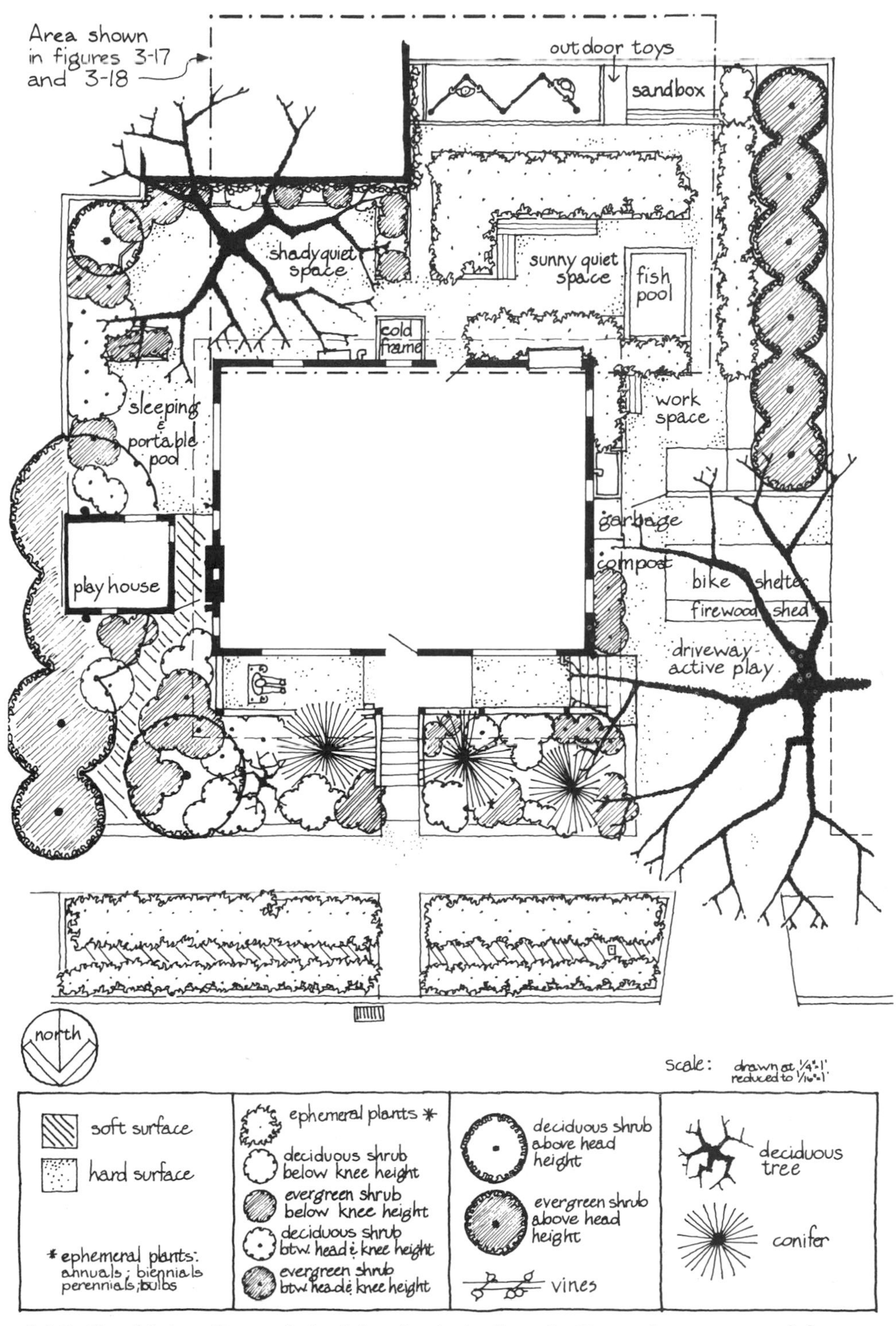


**2-15.** Combining Site and Activity Analysis: Step 2. General textures and forms. General types of ground surfaces (hard, soft) and plants (deciduous, evergreen) are added to the design, but no decisions about specific materials and plants are made yet.

2-16. Tulip and hardy bunching onion bloom together in spring.

2-18. Fennel shares space with 'Lulu' marigold.

2-17. Contrasting leaf textures of sage and burnet.

plan for a shrub mass of certain dimensions, which shrubs will fit these dimensions at maturity and thrive in your growing conditions? Are any of these plants edible? Make sure you choose those that are easy to care for and live with within the framework of your design. For example, choose a small fruit tree or commit yourself to pruning so you can reach the fruit on your edible shade tree; avoid plants plagued by disease in your climate. If you are committed to certain plants, adjust your design so the plants become easy to care for and live with. Do not, for example, plant an apple tree over a patio where the apples will drop on people's heads, and, if the apple tree is already growing on the site, move the patio to a different part of the design.

Once you narrow the field to easygoing plants well adapted to your garden, you still need to decide which plants look best to you, produce the largest quantities of food you are likely to use, and look well together. In plant combinations you can use a gradual sequence of leaf sizes and textures, or strong contrasts—large leaves with small ones, bold with fine textures, for example. (See figure 2-17.) Plant size is one aspect of landscaping that needs to be sequential. Abrupt contrasts of large and small plants render the small ones almost invisible, defeating your purpose.

You can combine food plants of similar leaf color and contrasting texture (blue-green cabbage and fennel, for example), or similar leaf color with contrasting bright ornamental flowers—gray-leaved sage, eggplant, and lagenaria gourd with rose pink cosmos, for example. Strong, clear greens, such as those of corn, celery, and parsley, look best with the brightest flowers—yellow, orange, red—and with pepper plants, which combine clear green foliage with colorful fruits. Combining similar foliage colors is suggested in Grant and Grant's *Garden Design Illustrated.* (See Bibliography.) The Grants have observed that plants of similar leaf color often look and grow well together because they are native to the same environmental conditions—gray-green plants to desert or dry environments, dark green plants to forests, bright yellow-green plants to wet or pondside environments. This idea is very useful, although there is no reason why you have to use similar foliage colors all the time. They can also be strongly contrasted.

Timing is very important. Some plants, such as late summer corn and sunflowers, and early lettuce and spring bulbs under flowering fruit trees, look best at the same time and complement each other. Other plants hide each other's weaknesses. Bushy zinnias conceal leeks during the summer, while the leeks grow slowly bigger; the zinnias obligingly die at first frost, as the leeks begin to look splendid. Timing varies with environment and plant variety. Try to use plants that are attractive for a long period of time. This way there will be some reliable overlapping in your plant combinations.

## Step 4. Positioning Food Plants in Garden Design

Exactly where will you put your food plants? They grow best in well-drained, loose-textured, fertile soil that is not excessively acid or alkaline. If you do not have this kind of soil already, you can improve the soil you have. Most food plants grow best in full sun, and tall ones should be planted on the north side of smaller plants so the small ones are not shaded.

Avoid putting food plants in soil where weed killer has been used, or where dog and cat excrement is a problem. Avoid places near heavy automobile traffic.

If their basic needs are met, food plants can be grown anywhere in your garden. Fruit and nut trees can be used as shade trees; some can be trained into flowering hedges. Fruit bushes can serve as low hedges and foundation plantings, or share border space with ornamental shrubs. Grapes, berry vines, cucumbers, squash, gourds, beans, and tomatoes can be trained on fences and trellises to provide privacy and shade. Strawberries and creeping herbs can serve as ground covers. Tall vegetables, such as corn, sunflowers, sorghum, and Jerusalem artichoke can form temporary hedges. Small vegetables and herbs can cluster around shrubs, move into flower borders, thrive densely in beds, and sprout out of containers. Annual vegetables and herbs should be planted next to deep-rooted plants that will not be disturbed by temporary company. Plants with fibrous, surface-feeding roots cannot tolerate much disturbance, and planting other things close around them may put them under stress. Give them room.

If you have special areas where food plants are dominant, you can plant annuals and perennials in separate parts of these areas, or, root systems permitting, mix them up. (See figure 2-18.) Vegetables and herbs can share their beds with annual and perennial flowers and spring flowering bulbs. If there is good, loose soil in these beds, there is no reason why you have to turn it every year! Soil turning is a habit left over from field farming. It is necessary if your soil is lumpy and you need to break it up into smaller pieces, or if you want to mix in large amounts of organic matter to improve soil texture. But if you have already been working on your soil and it is in good shape, you can replenish the soil as you go along, mixing in well-decomposed organic material near the soil surface, in various parts of beds whenever they are vacant. (Perhaps you would rather turn the soil over all at once, because it is easier and more clear-cut. This means, however, that you have to separate annual and perennial plants, and annuals that grow at different seasons.) In a mixed bed—mixed food plants and ornamentals, mixed annuals, biennials, and perennials—you can keep things growing and beautiful almost all year round. (See figures 2-19 and 2-20.)

**2-19.** Combining Site and Activity Analysis: Step 3. Specific textures and plants. Area of site in backyard along south wall of house in mid spring.

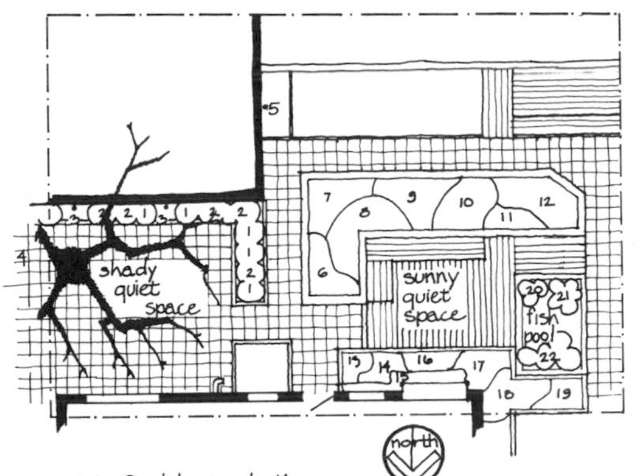


1. salal, Gaultheria shallon
2. western sword fern, Polystichum munitum
   – any bare patches between 1 & 2 bloom in spring with western trillium, Trillium ovatum, & avalanche lily, Erythronium grandiflorum
3. Hall's honeysuckle, Lonicera japonica 'Halliana'
4. existing sweet cherry & duke cherry – 'Black Republican' & 'Royal Ann' – on same tree
5. sweet autumn clematis, Clematis paniculata
6. leek 'Unique'
7. kale 'Dwarf Blue Scotch'
8. tulip 'the Bishop' (purple)
9. bunching onion 'Kujo Green', Allium fistulosum
10. hyacinth 'L' Innocence' (white)
11. burnet, Poterium sanguisorba
12. cabbage 'Early Jersey Wakefield'
13. silver thyme
14. blue squill 'Spring Beauty', Scilla sibirica
15. pea 'Mammoth Melting Sugar'
16. erect rosemary
17. lettuce 'Prizehead'
18. parsley 'Dark Green Italian' – interplanted w/ 'February Gold' daffodil
19. chervil, Anthriscus cerefolium
20. flowering rush, Butomis umbellatus
21. arrowhead, local strain of Sagittaria latifolia
22. American lotus, Nelumbo lutea

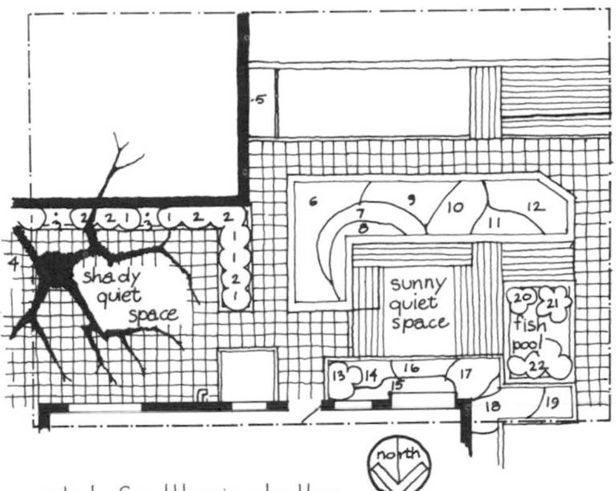


1. salal, Gaultheria shallon
2. western sword fern, Polystichum munitum
   any bare patches between 1 & 2 are filled during summer with erect fuschias
3. Hall's honeysuckle, Lonicera japonica 'Halliana'
4. existing cherry
5. sweet autumn clematis, Clematis paniculata
6. love-lies-bleeding 'Red', Amaranthus caudatus
7. snapdragon 'Rocket mixed'
8. soup celery 'French Dinant'
9. bunching onion 'Kujo Green', Allium fistulosum
10. cosmos 'White'/'Purity', Cosmos bipinnatus
11. Lilium speciosum var. rubrum
12. sweet fennel, Foeniculum vulgare
13. silver thyme
14. Chinese chives, Allium tuberosum
15. fuzzy gourd, Benincasa hispida, trellissed with morning glory 'Heavenly Blue'
16. erect rosemary
17. signet marigold 'Lulu', Tagetes signata pumila
18. hot pepper 'Red Chile'
19. basil 'Dark Opal'
20. flowering rush, Butomis umbellatus
21. arrowhead, local strain of Sagittaria latifolia
22. American lotus, Nelumbo lutea

**2-20.** Combining Site and Activity Analysis: Step 3. Specific textures and plants. Same area of site in late summer. Leek and burnet, from Figure 2-19, could have been left in place as perennials, but were instead moved to another part of the site.

# INEDIBLE ORNAMENTALS TO GROW WITH FOOD PLANTS

Inedible ornamentals can add a great deal of color and variety to your beautiful food garden. You may choose to plant a particular ornamental because you have a special fondness for it or because it is perfect for an environmental niche or landscape feature in your garden. You may want it for companion planting or for attracting wildlife; or, it may already be growing on your land.

Some ornamentals are better than others to grow near food plants. Avoid those, for example, that create dense shade or have voracious root systems. Roots should be fairly deep and sturdy, so that they will not be disturbed when annual vegetables and herbs are grown around them. They should be able to tolerate the plentiful amounts of water and soil nutrients that food plants require for good growth. Avoid ornamentals with insect and disease problems that require the use of heavy-duty chemical sprays which will contaminate nearby food plants. If small children sometimes play in your garden, do not grow poisonous ornamentals, especially those with poison berries, anywhere near food plants.

Ornamentals can be quiet-looking foliage plants that provide a pleasant-textured background for spectacular food plants, or showy plants themselves with several interesting features—bright flowers and fruit, appealing fragrance, fall leaf color, graceful branches, attractive bark. For maximum use of space in small gardens, the plants with many interesting features are preferable. Their attractive features should be as long-lasting as possible to make it easy to use other plants in combination with them.

If you want to grow food plants near a well-established ornamental, use an experimental approach. Dig into the soil near the ornamental. If you encounter an impenetrable root mass, you may as well give up. If not, start sturdy annual food plants nearby and make sure enough water and soil nutrients are available to satisfy all the plants involved. Under these conditions, the ornamental is unlikely to suffer any damage. However, watch food plants for low yield, yellow leaves, or other signs of stress. If the plants all thrive together in this first season, you can continue to put in annual food plants or switch to perennial herbs and fruit bushes.

## Ornamental Trees

Most cone-bearing evergreen trees create dense shade that food plants cannot tolerate. Some deciduous trees make dense shade, and trees such as poplars, willows, elms, and big maples have greedy roots. A tree's root system is often an approximate mirror image of its

2-21. Shade-tolerant combination of flower kale and impatiens. (Jim and Anita Green's garden, Corvallis, Oregon.)

branching system; roots stretch as far as or farther than branches do. Some trees have easily damaged roots (magnolia, for example), or roots that sucker prolifically when disturbed by cultivation (ailanthus, sassafras, zelkova). These trees are inconvenient to grow in close contact with food plants, but you can certainly try them if you haven't any choice.

Shade-tolerant fruit bushes, and annual and perennial vegetables and herbs (see figure 2-21) can most easily be grown beneath narrow-canopied or open-textured trees with deep, sturdy roots. A partial list of such trees includes albizia, Amur cork tree, honey locust, katsura tree, *Koelreuteria* species, Russian olive, sophora, styrax, tamarack, and dawn redwood. Many other trees are also good possibilities.

Suggested plant combinations are sophora and mint, honey locust and day lilies, and golden-rain tree and salal.

## Ornamental Shrubs

Many ornamental shrubs, both evergreen and deciduous, can easily coexist with food plants, especially with fruit bushes and shallow-rooted, edible ground covers. Many will also tolerate tree shade and the presence of herbaceous plants around them. Annuals can be planted in bays between shrubs, or a big space can be left in front of a shrub

2-22. Annuals—bush snap beans and marigolds—planted in front of privet, and abelia bushes. (Gertrude Mourer's garden, Eugene, Oregon.)

hedge for a flower-vegetable-herb border. (See figure 2-22.) Some shrubs can be clipped as edging plants for geometric beds of food plants. Shrubs with bright fall color can back up showy fall vegetables or grow in the shade of fruit trees with vivid autumn leaves. Bright-flowered shrubs can be planted near flowering fruit trees or herbaceous food plants that will look good at the same time they do. Cold-hardy vegetables and herbs can encircle evergreen shrubs.

Heath-family evergreen shrubs—rhododendron, azalea, mountain laurel, heath, heather, and pieris—have a strong need for acid soil, and fibrous root systems that do best when mulched and left undisturbed. They can be grown near blueberries, which have similar requirements. Desert shrubs should be grown only near drought-tolerant food plants, such as rosemary and sage. Shrubs with borderline cold-hardiness in your climate should be grown where they will not receive much water or fertilizer late in summer, since they need a chance to harden off.

Suggested plant combinations are serviceberry and winged euonymous (most effective in fall when leaves change color), witch hazel and kale (attractive in fall), dwarf pine and leek (especially attractive in winter), and santolina edging for geometric beds of sage, rosemary, or bunching onion (effective year-round in mild climates).



## Vines

Woody perennial vines can be trellised as backdrops for food plants;

vining food plants and ornamentals can grow together if they are of equal vigor or if the more rampant ones are controlled by pruning.

Suggested plant combinations are thornless blackberry and climbing rose, akebia and clematis, and grape and honeysuckle.

## Ground Covers

Ornamental ground covers can be mixed with edible ground covers of similar habit or the two can be grown in adjacent patches. They can be planted under fruit and nut trees and around the base of shrubs. Some ground covers—ivy and Saint Johnswort, for example—spread so vigorously that they cannot have any companions.

Suggested plant combinations are ajuga mixed with mint or violets, strawberry mixed with dwarf phlox, creeping thyme and prostrate juniper planted next to each other, and creeping potentilla under fruit trees.

## Herbaceous Perennials

Herbaceous perennials can be grown around fruit bushes. Those that tolerate shade—ferns, for example—thrive beneath the branches of fruit and nut trees. Perennial flowers can be mixed with perennial or annual vegetables and herbs. Some perennials have long flowering seasons, which makes them easy to combine with other plants. A partial list of long-flowering perennials includes beebalm, black-eyed Susan, campanula, chrysanthemum, coreopsis, platycedon, and Shasta daisy.

Suggested plant combinations are blueberry and Dutchman's breeches, soybean and black-eyed Susan, artichoke and Oriental poppy, and persimmon and sword ferns.

## Bulbs

Spring-flowering bulbs thrive beneath fruit trees, as well as at the base of fruit bushes, fruit vines, and espaliers. After mild winters, parsley, kale, and leek leaves are a beautiful setting for early snowdrops, crocuses, daffodils, and grape hyacinths. Late spring bulbs—irises, lilies, ornamental alliums—are in cheerful contrast to fresh green leaf vegetables sown in spring. After spring-flowering bulbs die down in summer, seeds of annual food plants can usually be sown in the same soil. Soil of very old bulb plantings may contain substances that inhibit the growth of other plants.

Suggested plant combinations are strawberries and irises, plum tree and daffodils, sweet cherry and tulips, and cornelian cherry and crocuses.

2-23. Snapdragon and celery thriving together.

## Annual Flowers

Annual flowers are extremely versatile. They fill temporary gaps between slow-growing fruit bushes and perennial herbs. Many thrive in poor soils and create bright spots of color in parts of the garden where little else will grow. Annual flowers can be sown at the same time as annual vegetables and herbs and cared for right along with them. (See figure 2-23.) In very moist, fertile soil, the flowers may bloom less prolifically. Many annual flowers, such as calliopsis, cosmos, marigold, pansy, salvia, scabiosa, snapdragon, zinnia, and others, have a long season of bloom. Those that bloom in spring and fall lend bright contrast to cool-weather vegetables. Tall flowers like tithonia and cleome help give fast, substantial presence to a garden. Annual flowering vines can be grown vertically with vining food plants. Many food plants have abundant foliage, useful for hiding the wispy or ugly leaves of some large-flowered annuals.

Some suggested plant combinations are zucchini squash and zinnia, corn and cleome, parsley and pansies, and currant bushes and marigolds.

In Chapter 3, entries on individual vegetables and herbs include suggestions for inedible ornamentals that can be grown with each of them.

## Chapter 3

# Encyclopedia of Attractive Vegetables and Herbs

People often think that attractive food plants mean only dwarf varieties or those with red or variegated leaves—red lettuce, 'Rhubarb' chard, and miniature pepper plants.

We do not have such restricted standards for inedible ornamentals. Ornamental garden landscapes would look very strange if all the plants in them were tiny or had red or variegated leaves. How *do* we judge traditional ornamentals? Important characteristics include plant growth habit; color, shape, and texture of leaves; prominence and beauty of flowers and fruits; and length of season that the plant is interesting and effective in the garden. A wide range of growth habits, leaf types, flowers, and fruits is considered attractive, although people have individual preferences within this range. Negative terms one hears most frequently are "coarse," "weedy," and "hard to grow." Of these, "hard to grow" is usually the most damning, although it is not a strictly aesthetic consideration.

You can judge food plants by the same standards as conventional ornamentals, but, at the same time, food plants must also have good yield and good flavor. Plenty of plants meet all these standards. They are attractive, they are easy to care for, and they make good food sources.

They must also be evaluated in terms of the particular situations where they will be growing. If the questions that follow are given careful thought, you can save a good deal of time and frustration.

**Environmental Considerations:** Will a given plant thrive in the soil, sunlight, and moisture conditions you can provide? Does it have reasonable insect and disease resistance in your garden environment? If it is a perennial, is it cold-hardy in your climate? Does the plant have any special characteristics it may need to grow and produce food in your climate, such as ability to germinate seed in cool or hot soil, to produce a good crop in very cool or hot conditions, etc.?

**Landscape Characteristics:** Do you like the way the plant looks? Are its growth habit, leaf type, eventual size, and other characteristics appropriate for its place and purpose in your garden?

**Personal Tastes:** Does the plant fit your individual/family style in gardening, cooking, and eating? For example, consider whether you want to grow a good many things to preserve for the winter, or whether you are unlikely to get around to it. Do you enjoy experimenting with new tastes, or would you rather stick to those you know you like? How much time do you want to spend taking care of your garden? Is this plant appropriate to your gardening habits?

## SOURCES OF ATTRACTIVE FOOD PLANTS

Attractive food plants from which you can choose are listed in this chapter and the next. Once you have decided which plants you want to grow, there are many sources available. Seed and nursery catalogs carry a wide variety of seeds and plants. Local nurseries and other stores sell seeds and plants, although selection is limited. Fruits, seeds, bulbs, corms, tubers, and roots from which you can grow food plants are sometimes sold as food in grocery stores.

A few plants, such as arrowhead, are available mostly from the wild. Before you collect them, make sure they are not on your state's endangered species list or growing in a protected area.

You can ask friends and neighbors for cuttings and propagate your own plants. Neighbors may have far too much of the very plant you want. They would be glad to dig up and give you as much as you will take!

As you look for new varieties to try, you may encounter some confusion. You will notice that food plant descriptions in seed catalogs are sometimes sketchy. For example, you can find thorough descriptions and great pictures of cabbage and lettuce, but it is not so easy to discover which carrots have vigorous, erect tops, as well as tasty roots. Photographs and descriptions lean heavily to roots alone, since that is the edible portion. You have to look for clues such as "sufficient top for good bunching" and "developed for a stronger top, which is essential on heavy ground where the tops break off easily when being pulled" (Stokes Seed Catalog, 1976), which can be translated as "vigorous erect tops." Tomato, eggplant, pepper, and squash descriptions usually focus heavily on fruits, and you must guess about the plants from any available hints. Pictures of whole plants are a great help when you can find them.

Two of the best sources of information are other people's gardens, and nearby variety trials or demonstration gardens sponsored by botanic gardens or the Cooperative Extension Service.

You may want to raise food from plants that are usually grown as ornamentals. It is often hard to track down information on good-tasting varieties and the conditions they require for good yield. You may find yourself in murky territory, trying to find recipes that use tiger lily bulbs or discover which bamboos have the best-flavored shoots. Possibilities to Explore at the end of the chapter lists some food uses of plants that are generally considered ornamentals. Books with further information about such plants include *Sturtevant's Edible Plants of the World* by E. L. Sturtevant and *Wyman's Gardening Encyclopedia* by Donald Wyman (see Bibliography). If the plant you are interested in is frequently used for food in another country, a gardener native to that country may be an excellent and delighted source of information. Botanic gardens and extension agents may be able to refer you to books or people that can help in the search.

If you go to a nursery to buy obscure food plants that are usually grown only as ornamentals, find out what pesticides have been used. Refer to a handbook of pesticides, such as *Farm Chemicals Handbook* edited by Gordon Berg (see Bibliography), to make sure the plant is or will be safe to use as food. If the chemicals that were used are both long-lasting and systemic (present throughout the plant's system), it is possible that even the fruits of the plant should not be eaten if they later appear. This depends on the chemicals involved. If safe plants are hard to find, ask the nursery to grow or obtain some for you.

## EVALUATING VEGETABLE AND HERB VARIETIES

Almost any plant, whether it is a food plant or a standard ornamental, has both strengths and weaknesses. You need to know about *all* its characteristics, not just the good ones, before you plant it in your garden. In this chapter, therefore, the advantages and disadvantages of each plant are listed in each entry.

It is important to have precise information on variety differences among food plants, because most food plants are not bred specifically for ornamental characteristics. With some food plants, almost all varieties are attractive; it is hard to go wrong. Others have no attractive varieties at all. The majority of food plants have a good deal of variability. For example, some eggplant varieties are attractive throughout summer, but others deteriorate at fruit-ripening time—branches

3-1. A sturdy carrot variety, 'Royal Chantenay,' used as an edging plant and backed up by marigolds.

break, and leaves may become browned with disease. Although some carrot varieties work very well as borders for paths, others have sparse or weak-stemmed foliage which trails along the ground.

The vegetables and herbs described in this chapter have been evaluated in terms of the characteristics listed here.

**Adaptability:** Some vegetables and herbs have exacting needs; others are very flexible. Most of them, however, need plenty of sun, adequate water, and soil that is fairly loose and well drained, well provided with nutrients, and not excessively acid or alkaline. Without these conditions they may grow, but they will not look as attractive or yield as well.

**Insect and Disease Resistance and Susceptibility:** There are many resistant vegetable varieties available. This is fortunate, because some insects and diseases lower or eliminate crop yield, and, even without significantly affecting yield, they can make plants look ugly. Other insects and diseases may be present but not important, or important only at certain times in the growing season. Many pungent herbs are not seriously affected by any insects or diseases.

**Flavor:** In order to avoid recommending plants that look good but taste bad, I tasted nearly everything I grew. I tried to be as broad-minded as possible, but some things tasted bad to me—excessive bitterness, toughness, stringiness, and peculiar off-tastes—and I think they would to most people. Some plants tasted strange but pleasant, and I have tried to describe their flavors so you can decide whether you want to try them.

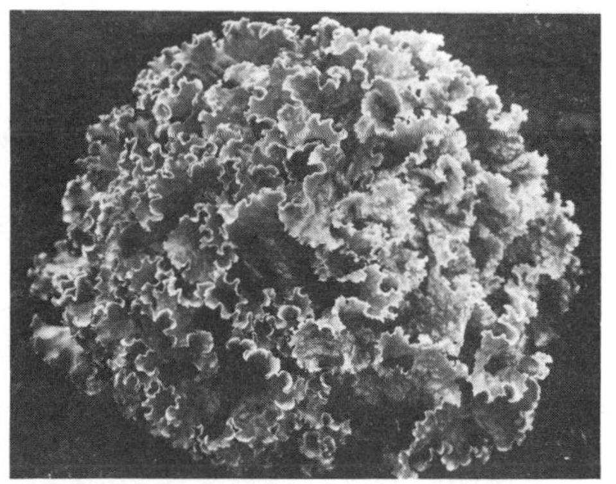

3-2a and b. Lettuce varieties with very different leaf textures: (a) 'Valmaine Cos' (Photo by George Gessert, Jr.); (b) 'Green Ice.'

**Yield:** Crop yield is defined here as yield per unit of space and time, not yield per plant. If a winter squash plant bears 10 fruits but takes 120 days until harvest and occupies 200 square feet (18 sq m) of space, that obviously limits its uses.

**Growth Habit:** Responses to plant growth habit are very personal. For two years, a panel of twenty other people worked with me on variety evaluation. They convinced me that not everyone shares my prejudices. As much as possible, I have tried to simply *describe* various growth habits, and I have included my value judgments only when I felt strongly. I like plants with compact, bushy growth and good vigor—plants that look natural, graceful, and well balanced. I do *not* like scrawny, leggy plants, and neither did the panelists. I do not like plants that are so dwarfed they look all pushed together, but many panelists disagreed.

**Leaf Color, Shape, and Texture:** Many of the panelists preferred frilly leaf textures to plain ones and variegated or nongreen colors to green. I have described various leaf textures, shapes, and colors, and you can choose among them. There is a great deal of variation: all-green leaves alone range from chartreuse and pale green, through gray-green, blue-green, and bright yellow-green, to dark green, with a wide range of shapes and textures. A broad choice of effects may be available with a single vegetable, such as lettuce (see figures 3-2a and b), because its varieties are so different from each other.

**Attractiveness and Prominence of Flowers and Fruits:** Few vegetable and herb flowers are actually ugly, but some are inconspicuous. This is important only if flowers and fruits are present during a plant's life cycle in your garden. Beet flowers, for example, do not appear until the plants' second year; by then plants have been harvested.

**Length of Season of Usefulness:** This is important to consider with regard to the plant both as a food source and as an attractive part of the landscape. The ideal attractive food plant would be beautiful in leaf, in flower, and in fruit. It would taste delicious and have a long season of harvest. It would be perennial and effective all winter. Few food plants meet all these conditions, but some of them satisfy many.

A plant that supplies food briefly, is attractive for only a short time, or has a short life cycle, so that it will have to be resown frequently, is less useful in edible landscaping than is a plant with more staying power. Plants of short-term value have a definite place in the garden, but you cannot rely heavily on them. Lettuce, for example, will not last more than a couple of months and will need to be replanted, or planted near something that will spread and cover the place where it has been removed. Love-lies-bleeding, a type of amaranthus, has delicious leaves in all stages of growth, but is not attractive until it flowers. With information about the limitations of plants like these, you can plan your garden successfully.

The vegetables and herbs that appear in this chapter are the best I have found so far in terms of all the characteristics described above. Only the most and least attractive varieties are described; other varieties tested appear in Appendix II. Plant variety names are enclosed in single quotation marks: for example, 'Autumn King' carrot. These varieties (which are also called horticultural varieties or cultivars) have been bred and selected by plant breeders. These varieties are different from "botanical varieties," naturally occurring subdivisions of a species, which are described by Latin genus and species names followed by the abbreviation "var." and a third name: for example, *Amaranthus tricolor* var. *salicifolius.* A "subspecies" is a larger naturally occurring subdivision of a species than a botanical variety: for example, fennel, *Foeniculum vulgare* subspecies *vulgare.* A "group" is a group of similar horticultural varieties within a species, for example, cabbage, *Brassica Oleracea* Capitata Group.

Each vegetable and herb in the encyclopedia is listed under its most frequently used common name, followed by its Latin name. Plants with three stars (* * *) by their names are outstanding ornamental food plants: one-star (*) plants are pleasant but unspectacular, and those with two stars (* *) fall somewhere in between. Vegetables and herbs described as "hardy" survive frost to varying degrees. Seeds can be sown in early or midspring, depending on the climate and the plants' exact amount of hardiness. "Tender" plants do not survive frost; seedlings should be started early indoors, or seeds sown outdoors when there is no danger of frost. "Half-hardy" plants can tolerate some cold, but no hard frosts. Seeds can be sown outdoors after hard frosts are past.



Information in the encyclopedia is national in scope. You will need local information about planting dates, locally successful plant

3-3. Amaranthus love-lies-bleeding.

varieties, locally important insects and diseases, and the controls available for them. Good sources of information include the Cooperative Extension Service in your county, botanic gardens, garden clubs, neighborhood gardeners, and standard vegetable gardening books.

When growing vegetables in beds rather than rows (as is often suggested in the encyclopedia), plant them as far apart as they are likely to grow at maturity, or space them closer if you can eat the thinnings. Note: Plant sizes listed here are larger than plants will grow in most gardens. The cool weather and moist, fertile soil of our trial gardens encouraged large dimensions.

## ***Amaranthus (several *Amaranthus* species) tender annuals

**Ornamental Characteristics:** Several species of amaranthus are both attractive and edible and there are many varieties. They are bold, vigorous plants, with large, ovate leaves that may be green, red-brown, scarlet, or variegated. Most varieties stay attractive for a long time. Foliage types, usually varieties of Joseph's coat, are often brightly colored from the time they sprout. Leaves become especially colorful in late summer, although flowers may be somewhat inconspicuous.

Love-lies-bleeding, (*A. caudatus*) has plain green leaves, but when long, drooping red tassel flowers appear, the plant acquires a droll grace all its own and becomes an excellent ornamental. The flowers last several months.

**Ornamental Drawbacks:** When plants are small, the leaves of love-lies-bleeding resemble those of its relative, pigweed, a common weed.

**Adaptability:** Attractive amaranthus species have the weedy strength of pigweeds. They tolerate drought, heat, and poor soil, but need plenty of sun. Pinching growing tips when plants are small encourages bushiness and keeps them erect. They can be sown early indoors, or outdoors after the soil warms up. Sometimes they reseed themselves.

**Insects and Diseases:** Cucumber beetles and other insects sometimes eat holes in amaranthus leaves.

**Food Uses:** Vitamin-rich leaves are the edible portion of many amaranthus varieties. They are most tender when plants are young, but plants *look* best when they are older. You can sow thickly and harvest young thinnings; later in the summer you can eat new leaves that older plants produce. You can use amaranthus fresh or frozen, steamed, boiled, stir-fried, and added to soups. It is popular in China and Japan, and in the East and West Indies.

Several wild amaranthus species contain high quantities of nitrates in their leaves; it may be best to eat all amaranthus leaves in moderation.

Some amaranthus varieties are grown for edible seed that can be ground into flour. Yield is high and seeds are relatively easy to collect and clean. Their protein content is outstanding.

**Landscape Uses:** Amaranthus is popular as an ornamental in the United States, although it is seldom eaten here. Amaranthus usually appears in flower sections of seed catalogs. It looks good against somber, monochromatic backgrounds, such as gray walls or evergreen hedges. It can be combined with other plants that are big and bold—eggplant, 'Rhubarb' chard, sorghum, squash, red-brown corn—and with other varieties of amaranthus. Delicate-textured plants look good with amaranthus if they are large, like fennel. Amaranthus and celosia are often grown together in flower borders, but I think they make an odd, garish combination. Sesame and love-lies-bleeding are excellent companions; so are 'Dark Opal' basil and amaranthus 'Molten Fire,' with leaves of almost the same color.

**Species and Varieties:** Leaves of Joseph's coat (*A. tricolor* 'Splendens' = *A. gangeticus*) are usually green with red and yellow splotches. The varieties 'Molten Fire' and 'Early Splendor' have bronze leaves variegated with maroon. These varieties are 1 to 4 feet tall (.3 to 1.2 m) and are sometimes smaller-leaved than other amaranthus. Tampala

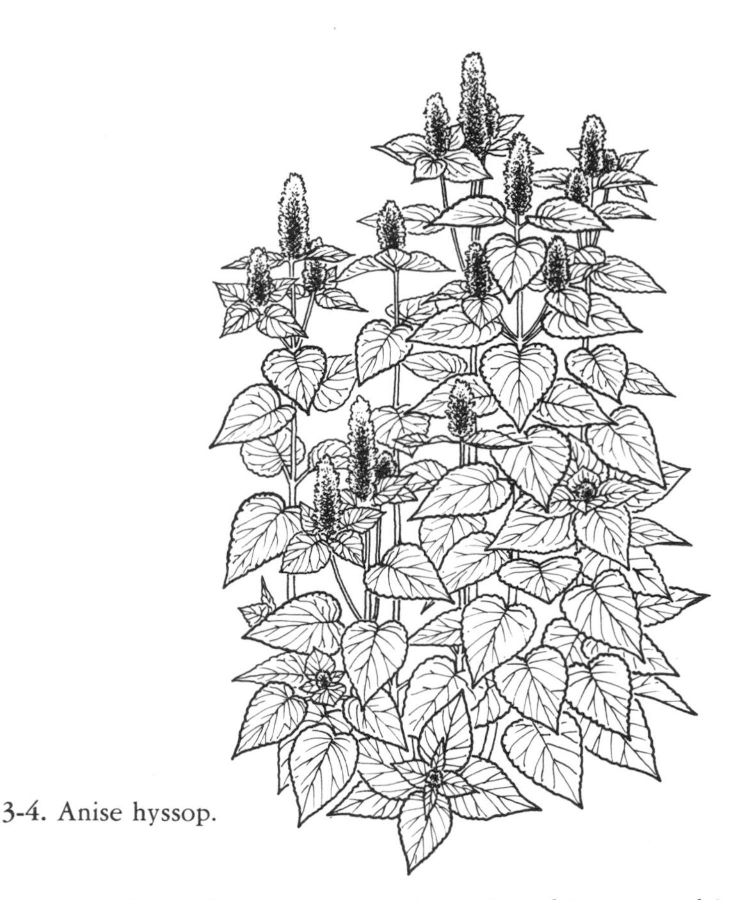

3-4. Anise hyssop.

is an unattractive, green-leaved variety sometimes listed in vegetable catalogs.

Love-lies-bleeding ordinarily has green leaves and red tassel flowers, but some forms have red-brown leaves, and white, yellow, green, or pink tassels. Plants are tall and spreading—up to 4 feet tall and 5 feet wide (1.2 by 1.5 m).

Many other varieties of Joseph's coat and love-lies-bleeding deserve trial. So do narrow-leaved *A. tricolor* var. *salicifolius,* and *A. hybridus* var. *erythrostachys,* prince's feather. Some varieties of prince's feather have bronze foliage and attractive flower spikes; some have edible leaves and a high yield of edible seeds.

## ***Anise Hyssop (*Agastache foeniculum*)
hardy perennial

**Ornamental Characteristics:** Anise hyssop's green leaves are heart-shaped, with strongly scalloped edges and a delicate surface tracery of veins. Its growth habit is sturdy, erect, and unusually narrow. It grows to 20 inches tall and 1 foot across (50 by 30 cm). It blooms for three months, with a resting period in the middle before a new flush of bloom begins. The top of each plant is covered with dense bluish purple flower spikes loaded with nectar. They swarm with honeybees and bumblebees in the daytime, and with moths in the evening.

**Ornamental Drawbacks:** I know of none. It is a beautiful plant.
**Adaptability:** Anise hyssop can be sown outdoors in fall or early spring, or started indoors. It may grow slowly at first and blooms about two months from seed. It dies back to the ground in fall and sprouts again the next spring.
**Insects and Diseases:** Anise hyssop is very healthy.
**Food Uses:** Leaves smell and taste like anise or licorice, but are slightly more mentholated. They can be used for seasoning, fresh or dried. Plains Indians brewed them for tea. Birds like the seeds.
**Landscape Uses:** Anise hyssop makes an excellent formal edging for a path, either alone or with a front row of shorter plants, such as parsley or dwarf marigolds. It can be grown in clumps near other plants with blue-purple, blue, or blue-green colors—ageratum, heliotrope, blue salvia, hyssop, sage, lavender, cornflower, rue, and fennel. Its flowers also look excellent with clear yellow, white, or rose flowers. Other strongly erect plants, such as hollyhock and blue salvia, repeat and increase its vertical effect.
**Species:** The leaves of several other *Agastache* species are used for seasoning and tea in other countries. Mexican giant hyssop (*A. mexicana*) has big pink flower spikes. Korean mint (*A. rugosa*) is an anise-scented plant grown in Oriental countries.

## **Arrowhead (several *Sagittaria* species)
hardy perennial aquatics

**Ornamental Characteristics:** Arrowheads live in shallow water or mud, in swamps, and along the edges of streams, ponds, and lakes. Shiny green leaves shaped like arrow tips rise on upright stems. Three-petaled white flowers with yellow centers bloom through the summer; they are followed by clusters of bumpy seedpods. Arrowhead plants are 6 to 36 inches (15 to 90 cm) tall and spread rapidly as underground stems send up new shoots.
**Ornamental Drawbacks:** Some species grow more rampantly than others and can be a problem in small pools. Even if they are planted in roomy pots, they may still escape and send up sprouts.
**Adaptability:** Arrowheads can be started in spring from tubers or division and are easily transplanted. They thrive in rich soil and should be kept constantly wet or submerged in up to 6 inches (15 cm) of water.

Leaves die in the winter; roots and tubers are quite hardy.
**Insects and Diseases:** Aphids can be numerous and very difficult to get rid of.



**Food Uses:** You can harvest arrowhead tubers, borne at the end of the roots, through fall and winter. They were an important starchy staple for some American Indian tribes, and are often eaten in China

3-5. Arrowhead.

and Japan today. Roast them, boil them, and use them in stir-fries and stews. Their pleasant flavor is between that of a potato and a sweet, mild nut.

**Landscape Uses:** Arrowheads can grow in ponds, washtubs, barrels, or pots. Plants look best in a dense mass, but produce most heavily if they are at least 12 inches (30 cm) apart.

Arrowheads look attractive with upright water plants, such as water chestnut, cattail, and pickerel weed, and low, contrasting plants, such as water lily. The goldfish hide among them from cats and other predators.

Tall forms of arrowhead may rise from the water and tower above plants on land; small arrowheads may be lower than shore plants. White petunias, geraniums, and impatiens are attractive growing nearby on dry land; neat-leaved 'Delicata' squash is a harmonious background.

**Species and Varieties:** Some varieties of Chinese arrowhead (*S. sagittifolia*) are double-flowered, and all have a purple area at the base of each petal. Chinese arrowheads grown from tubers may not flower during their first summer, and roots do not spread rampantly.

Arrowhead or wapato *(S. latifolia),* native across North America, is a variable species. It spreads quickly and flowers the same summer that tubers are planted.

Both species have good-sized tubers. Other species grow locally in North America; giant arrowhead *(S. montevidensis)* is native to the southern part of the continent.

Arrowhead tubers can easily be collected from the wild, but make sure that the plants you collect are not on your state's endangered list

3-6. Globe artichoke.

and that they are not growing in a protected area. Locally collected plants are most likely to thrive, but often may not produce the biggest tubers.

Chinese arrowhead tubers are sometimes sold at Oriental groceries. Giant-leaved and double-flowered arrowheads are available from water plant nurseries.

## **Artichoke, Globe (*Cynara scolymus*)
hardy perennial

**Ornamental Characteristics:** Globe artichokes are dramatic-looking plants that form open rosettes up to 6 feet tall and 6 feet across (1.8 by 1.8 m). The pale gray green leaves have deep, pointed lobes; young leaves are fuzzy and older ones have a somewhat shiny surface. Each season six or more large flower heads appear on each plant. Layers of bracts enclose the flower buds. If the heads are not eaten, they open into big, dramatic, purple flowers like those of thistles. Artichokes may produce through fall or until hard frost. The leaves last through mild winters.
**Ornamental Drawbacks:** Artichokes are too big to fit in every garden. Their jagged leaves seem weedy to some people.
**Adaptability:** Plants are usually propagated by dividing the offshoots they produce each year. They can also be grown from seed, if they are started early indoors. The first year from seed, artichokes produce flower buds in late summer or early fall; during successive years, they produce buds in spring. If the spring buds are cut off, they produce again in fall.

Artichokes thrive in cool weather and with plenty of water and nutrients. Hot weather may cause the buds to get tough and open fast. Leaves are hardy to about 20°F. (-7°C.) and roots to somewhat lower temperatures. Various methods of winter protection are mulching, tying the leaves together over the crown, or storing the roots in a cold frame.

**Insects and Diseases:** Insect problems include aphids, slugs, snails, earwigs, and artichoke plume moths. Leaf spot and several other diseases are possible.

**Food Uses:** The fleshy, leaflike bracts, stem, and receptacle (base) of the artichoke's flower bud are its edible parts. Artichokes are usually boiled whole, and the bracts are peeled off and dipped in sauce. You can also stuff them.

The term "artichoke hearts" can refer to tiny whole buds or to receptacles of large buds. Hearts can be pickled or cooked fresh.

**Landscape Uses:** A row of artichokes makes a great summer hedge or garden space divider; the plants are striking in front of a dark wall or evergreen hedge.

A massed bed of artichokes is too bulky for most gardens. A single plant in a tub makes a space-consuming, but handsome, focal point for a patio.

The gray-green foliage of cardoon, sage, dusty miller, olive, and Russian olive matches that of artichokes. Big flowers are good companions: strong-colored zinnias and black-eyed Susans, and delicate rose shades of cleome.

**Varieties:** 'Green Globe', the only variety widely available in the United States, is valuable because it produces uniform plants from seed. However, several other varieties exist—'Purple Globe', 'Creole', and 'Vert de Laon'—and are all well worth trying, especially if you can get plants rather than seeds.

## ***Asparagus (*Asparagus officinalis*)
hardy perennial

**Ornamental Characteristics:** Sturdy asparagus spears push through the soil in early spring. Spears that are not harvested grow into 5-foot (1.5-m) masses of feathery, emerald-green foliage. The plants are delightful to look at during summer and fall. The small, white flowers are followed by bright red berries on female plants, and foliage turns clear yellow in fall. In winter the berries fall off and the plants turn brown.

**Ornamental Drawbacks:** Asparagus plants demand a long-term commitment. Well-maintained beds may last from ten to fifty years or more, but it takes awhile to get them going. If you plant asparagus from seed, you will have little to look at the first year and nothing to eat until the third year.

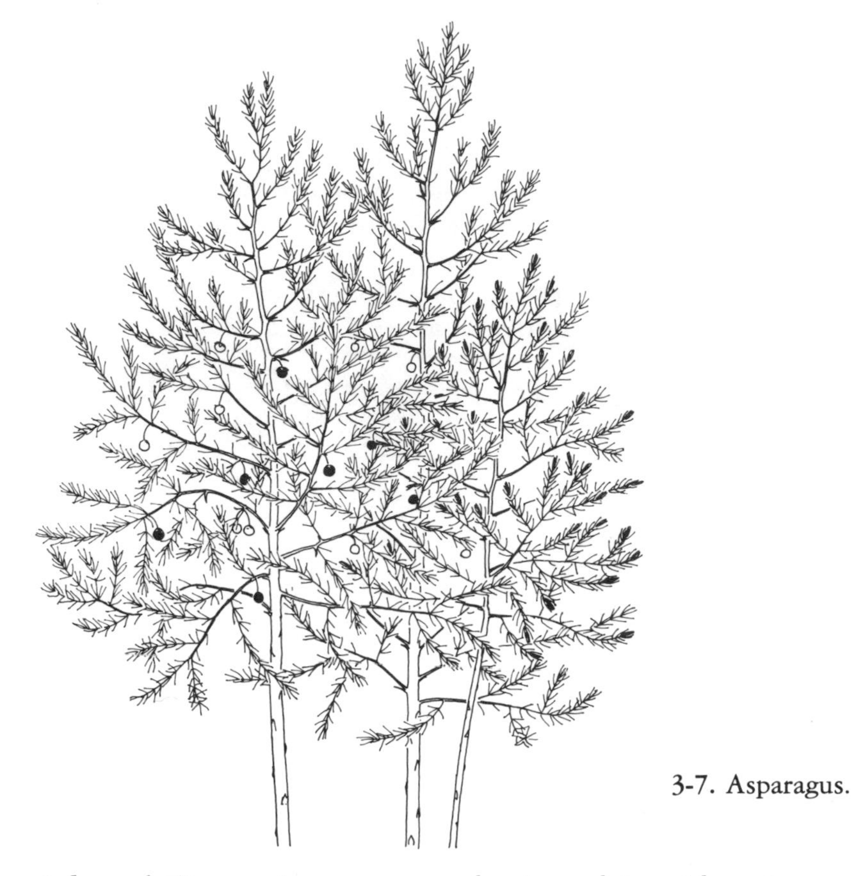

3-7. Asparagus.

**Adaptability:** Asparagus seeds planted in mid-spring germinate in about one month, and seedlings grow slowly the first summer. It saves time to buy and plant one-year-old dormant crowns. One method of planting is to lay the crowns 18 to 24 inches (45 to 60 cm) apart in 6- to 8-inch-deep (15- to 20-cm) holes or trenches. Cover the crowns with several inches of soil, gradually adding more as buds sprout and grow. Spears should not be harvested the year the crowns are planted. Most sources suggest a light harvest the second year and about two months' harvest in succeeding years. Some spears are left to develop and produce food that can be stored for the next year. Asparagus tolerates a wide range of climates, and all but heavy clay soil. It does best with a long, sunny growing season.

**Insects and Diseases:** The asparagus beetle is the most destructive insect. Principal diseases are rust and fusarium wilt. Rust-resistant varieties are widely available.

**Food Uses:** Use asparagus fresh, canned, or frozen. It is excellent in Oriental stir-fries as well as European dishes.

**Landscape Uses:** Asparagus makes a light, beautiful hedge, or an airy backdrop for a border. It is an attractive background for most herbaceous plants—those with fine textures, such as carrot, parsley, cosmos, and signet marigold, as well as the contrasting bold textures of zucchini squash or zinnia, and the spirelike flowers of hollyhock,

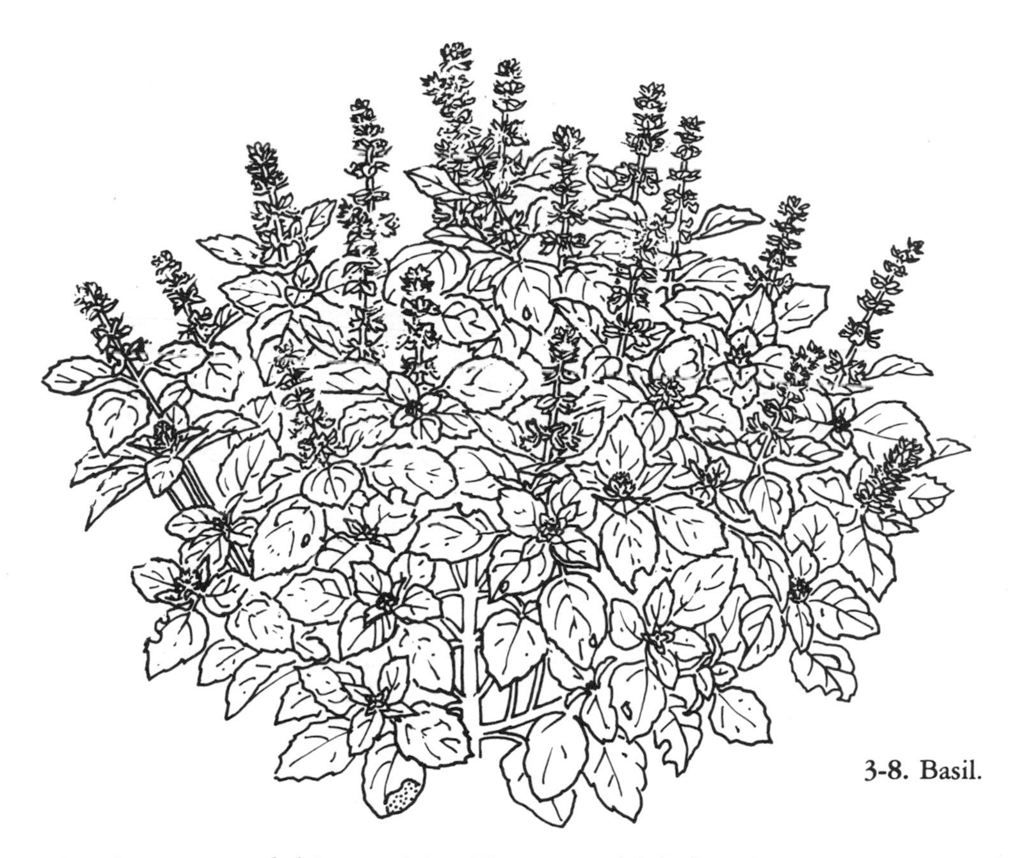
3-8. Basil.

anise hyssop, and blue salvia. Flowers which last into autumn are blooming when asparagus is most colorful; blue salvia and red or bronze chrysanthemums are excellent fall companions. Asparagus itself is lovely in lacy silhouette against a dark wall or evergreen foliage.
**Species and Varieties:** Strains of asparagus have been developed for specific parts of the country. 'Mary Washington' is a vigorous, widely available variety with good rust resistance. Foragers collect spears of wild *A. officinalis* growing along roadsides. Six other species are used in various parts of the world for their edible shoots.

## ****Basil (*Ocimum basilicum*)**
tender annual

**Ornamental Characteristics:** 'Dark Opal' basil and little-leaf basil are good ornamentals. The leaves of 'Dark Opal' (a variety of *O. basilicum*) are shiny, dark, reddish purple, smooth-surfaced, and rather small, with toothed edges. If you brush against them, a sweet, spicy fragrance is released. Plants are dense, neat bushes, 1½ feet (45 cm) tall and 2 feet (60 cm) wide by the time they are in flower. Small, lavender flower spikes are in delicate contrast to the dark purple leaves. Little-leaf basil (*O. basilicum 'Minimum'*) has tiny, light green leaves and white flower spikes. It forms dense mounds up to 1 foot (30 cm) tall and 1½ feet (45 cm) wide.

**Ornamental Drawbacks:** Some basil varieties are unattractive. (See "Species and Varieties.")
**Adaptability:** Basil grows best when days and nights are warm. It needs very fertile, well-aerated soil and plenty of moisture. Because outdoor germination may be spotty, it is advisable to start seeds indoors, six weeks early, and to move seedlings outside when the weather is warm.

Some books suggest pinching off flowers if the plants begin to bloom. This works for a week or two in the beginning, but the basil plants keep on setting more and more flowers. It is best to harvest when bloom begins. Cut plants back to 1 or 2 inches (2.5 to 5 cm) above the ground. They will grow back and produce a second crop.
**Insects and Diseases:** Cucumber beetles, Japanese beetles, slugs, and cutworms are possible problems.
**Food Uses:** Grind or blend fresh leaves with olive oil, garlic, and sometimes pine nuts to make pesto. Pesto, which can be frozen until needed, is mixed with parmesan cheese and used as sauce for noodles, vegetables, and poultry.

Chopped basil leaves taste good in salads. Fresh or frozen, they are excellent seasoning for a wide variety of cooked foods. They taste like they smell—warm and spicy. When dried, however, much of their flavor is lost.

You can make basil vinegar. Leaves of 'Dark Opal' basil are used to turn vinegar a ruby-red color (although they turn pesto a mud brown). In the Near East, basil seeds are eaten alone or cooked in breads.
**Landscape Uses:** Basil looks good massed in beds and borders, edging paths, or growing in pots and windowboxes. 'Dark Opal' basil contrasts well with clear yellow flowers or with the bright, strong green of parsley. Pink and lavender flowers are a close, harmonious combination. 'Dark Opal' basil's leaves are hard to match exactly with other leaves or flowers, but amaranthus 'Molten Fire' and varieties of Japanese maple come close. Little-leaf basil can be grown with fine-leaved plants—parsley and carrots—and with small, bright flowers—dwarf phlox and signet marigold.
**Species and Varieties:** Standard sweet basil *(O. basilicum)* has large yellow-green, rather blistered leaves. It is attractive when small, and its big leaves are useful for pesto, but it grows into a loose, spreading bush that looks untidy and weedy by the time it is in flower. Lettuce-leaf basil has twisted, heavily blistered, yellow-green leaves that look contorted. Other varieties and species of basil—sacred basil, tarragon-scented basil, clove-scented basil, tricolor basil, tree basil (a perennial shrub), and lemon basil, with its large, delightfully fragrant leaves and burgundy flowers—have good possibilities as ornamentals and food plants.

3-9. Bush snap bean.

## **Bean, Bush Snap (*Phaseolus vulgaris* var. *humilis*)
tender annual

**Ornamental Characteristics:** Although bush beans have limited ornamental possibilities, many gardeners grow them for food and because they are especially useful for urban gardens with high lead content in the soil. Fruiting vegetables have less lead in their edible portion than do root or leaf vegetables, but some of them do require a good deal of space and time to produce a crop, and many have conspicuous fruits that encourage vandalism. Bush beans, however, are small and early yielding; their fruits are often inconspicuous.

Some bush beans have an erect growth habit. Plants grow up to 1½ feet tall and 2½ feet wide (45 by 75 cm). Their large, medium-green leaflets are in groups of three. Some varieties bear yellow or purple pods in contrast to the leaves.

**Ornamental Drawbacks:** Many varieties are likely to fall over; many have blistered and somewhat twisted leaves. The small lavender or white flowers are often contorted.

**Adaptability:** Bean seeds germinate poorly in very cool soil. In many areas, they can be planted from mid-spring to early summer. Plants grow fast. Although they are legumes, they need plenty of nitrogen in the soil; lower leaves turn yellow when they do not get it.

**Insects and Diseases:** Bean diseases include viruses, rust, bacterial diseases, anthracnose, and others. Controls include the use of resistant varieties, disease-free seed, and chemical seed treatments. Insects encountered most frequently are aphids and Mexican bean beetles. Despite all these possibilities, beans are often very healthy.
**Food Uses:** Can, freeze, and fresh-cook snap beans. Stagger plantings or plant varieties with different maturity dates unless you want a lot of beans at once for freezing or canning.

Most modern snap beans do not develop strings in their pods even as seeds begin to swell. Some varieties are slightly juicier, sweeter, or meatier than others, but differences are very slight.
**Landscape Uses:** Dense plantings, 12 to 18 inches apart (30 to 45 cm), help plants stay upright. Make sure beds are small enough for easy harvest; if you pull hard on the pods, it is easy to uproot the plants.

Bush beans are good edging plants. You might contrast their foliage with the pale colors and delicate textures of baby's breath, pimpernel, and coriander. Beans with yellow or purple pods can be grown near flowers of similar color.
**Varieties:** 'Tendercrop' and 'Topcrop' are especially vigorous and attractive green-podded snap bean varieties. 'Cherokee Wax' is a good-tasting yellow-podded wax bean. 'Royal Burgundy' is a better-looking purple-podded variety than 'Royalty Purple Pod.' 'Royal Burgundy' has higher yield and more erect growth. Both varieties have a rather open, inferior growth habit compared to the most vigorous green-podded varieties.

Several dry bean varieties—'Pinto,' 'California Red Kidney,' and others—do not show promise as ornamentals. At first, they grow in bushes, but later, runners stretch in various directions. Vigorous 'Montezuma Red' was the most attractive variety in our trial gardens.

Dry beans are excellent food crops, but they have a long season until harvest. Plants must stay in the garden while seeds ripen and leaves turn brown and die.

## ** Bean, Runner and Pole (*Phaseolus coccineus* and *P. vulgaris*)

tender, mostly annual, *P. coccineus* is grown as a perennial in the tropics

**Ornamental Characteristics:** The climbing habit of pole and runner beans takes advantage of vertical space. Plants twine their stems around anything available and climb 10 feet (3 m) or more. Runner bean (*P. coccineus*) is the most attractive climbing bean. (See figure 3-10.) Its leaves are typically bean-shaped, but smooth-surfaced and rather bright green. Its flowers, larger and more conspicuous than

**3-10.** Runner Bean. (From Mm. Vilmorin-Andrieux's *The Vegetable Garden.*)

those of other beans, are purple, pink, red and white, or bright scarlet. However, runner bean is a less determined climber than other pole beans, twining rather loosely. It is often less densely leaved.

**Ornamental Drawbacks:** Like bush beans, pole beans have medium-green, blistered leaves and inconspicuous flowers. Leaves of all climbing beans may become yellow or brown because of cultural conditions and/or disease.

**Adaptability:** Pole and runner beans require a longer season than do bush snap beans to begin producing a crop, but they produce for a longer period of time. Like bush beans, they need plenty of nitrogen in the soil, and seeds germinate poorly in cool soil. You can plant them from mid-spring to late spring.

**Insects and Diseases:** The insects and diseases that affect bush beans affect runner and pole beans as well.

**Food Uses:** Some pole beans are grown for dry beans, but most are grown for their immature pods, used fresh, canned, or frozen. Some of the pods become stringy as they mature. The texture of runner bean pods is grainy and crunchy, and, as they grow larger, their flavor becomes strong. Some people find the strong flavor objectionable. Other pole beans taste just like bush snap beans. Like bush beans, climbing beans are excellent crops for urban gardens that have high lead content in the soil.

3-11. A bean-walled shelter makes a great summer house. (Joseph Rogers Gessert in the Lincoln Community School Garden, Eugene, Oregon.)

**Landscape Uses:** In traditional vegetable gardens, climbing beans often twine up vertical strings attached to a wire stretched between two posts. This system works well, but the beans are often stuck out in the middle of the garden, in a random place where they serve no landscaping purpose. Actually, there are many places in the garden where a bean wall is very useful. You can trellis beans along the side of a patio to give privacy or shade. You can grow them on a warm house wall to help with climate control inside the house. A wall with vegetation on it stays cooler than a bare wall, and, in cool climates, beans can use the extra heat the wall reflects. They can be grown on fences and laundry line posts. The Indians grew them on corn plants, but this arrangement can be hard on the corn. A bean tepee or a shelter with beans for walls makes a great summer house for children and adults. (See figure 3-11.)

If you like jungle-like profusion, grow beans vertically with cucumbers, wax gourds, morning glories, and other flowering vines. Allow 1 foot (30 cm) of soil space between plants so they have adequate water and soil nutrients and so you have a chance of finding fruits when they develop.

Other red flowers nearby—cardinal climber, red-flowered morning glory, and zinnia—emphasize the bright red blossoms of scarlet runner bean.



**Species and Varieties:** *P. vulgaris* pole beans include 'Kentucky Wonder,' 'Pole Blue Lake,' and 'Romano.' Their main ornamental contributions are vigorous leafiness and a strong climbing habit. They are

3-12. Beet.

best used for the mass effect of dense foliage and twining stems. 'Purple-Podded Pole Bean' has conspicuous pods, but is quite ugly in other ways.

Runner bean (*P. coccineus*) includes 'Scarlet Runner,' pink-flowered 'Sunset,' and red and white 'Painted Lady.' 'Scarlet Runner,' listed in vegetable or flower sections of U.S. seed catalogs, is the most vigorous, floriferous runner bean we have tested. Runner bean is a popular food plant in England and its other varieties are found mostly in English catalogs.

## *Beet (*Beta vulgaris,* Crassa Group)
hardy biennial

**Ornamental Characteristics:** Beets are not especially attractive, but a few have excellent fall color. Most beet leaves change from shiny medium green to an assortment of reds, greens, and yellows. The best varieties become deep crimson. Beets tolerate moderate frost.
**Ornamental Drawbacks:** In summer, the leaves are not showy. They are blistered and wavy-edged, with some red veining. Diseases often cause spots and dead patches.
**Adaptability:** Beets grow in a wide range of conditions and can be planted from early spring through summer. Roots mature in about two months so, for fall color, it is best to plant seeds in July or August. Roots taste best when small, so don't plant too early.

**Insects and Diseases:** Leaf disease susceptibilities are scab, downy mildew, powdery mildew, rust, and several leaf spots.
**Food Uses:** Use beet roots fresh, canned, pickled, or frozen. Store them in the garden or in a cool basement or root cellar. The leaves, rich in vitamins and minerals, make delicious cooked greens.
**Landscape Uses:** Beet tops are up to 15 inches high and 28 inches wide (38 by 70 cm). They may look odd near big plants, unless medium-sized plants are used as intermediaries.

In summer, you can plant beets behind something taller that will need to be pulled out early in fall, when the beets become showy. They look good near fall asters and chrysanthemums and can echo fall colors of deciduous trees and shrubs.
**Varieties:** The strongest and most uniform fall color appears in 'Early Red Ball' beets. 'Tendersweet Formanova,' a cylindrical-rooted variety developed for slicing, is almost as good. These varieties are sometimes affected by leaf diseases, but their resistance seems greater than that of many other varieties.

'Redpack' has outstanding health and vigor, but little fall color. Its leaves are somewhat larger than those of other beets, and red veining is more prominent. It looks very attractive near 'Rhubarb' chard.

## *Borage (*Borago officinalis*)
hardy annual

**Ornamental Characteristics:** At first, borage is a homely rosette of large, gray-green leaves covered with bumps and hairs. As it grows, it becomes wide at the base and narrow at the top until the leaf mass is pyramidal. Within two months from seed, hairy, erect stems rise up covered with dainty buds and flowers. Cinderella! The drooping, hairy buds are delicate rose purple. The nodding flowers have five sky-blue petals that bend slightly back and alternate with pointed, black sepals. Black stamens project outward in a narrow cone. Borage flowers resemble shooting stars, the woodland wildflowers. The plants bloom steadily and prolifically for several months. Fortunately, the plentiful, wilted flowers resemble the buds and do not turn brown and dead-looking until near the end of summer.
**Ornamental Drawbacks:** The leaves are not especially attractive. The bottom ones sometimes yellow after the plants have been blooming a while. Although plants are pyramidal as they begin flowering, they soon branch repeatedly and sprawl over a wide area in dense, bushy masses up to 3 feet wide and 2½ feet tall (90 by 75 cm).



**Adaptability:** Borage is extremely vigorous. It reseeds itself abundantly and new plants grow as winter begins. In mild climates the new plants overwinter and bloom early the next spring. Hard frost may nip

3-13. Borage.

stem tips and flowers, but plants recover and go on blooming. They do well in full sun, and in partial shade, where their growth habit is more delicate. They flourish in cool and warm summers. You can plant seeds from early spring through summer.

**Insects and Diseases:** Cucumber beetles, flea beetles, slugs, and Japanese beetles are possible insect problems.

**Food Uses:** Leaves and flowers have a mild, cucumber-like flavor. Young leaves are best used fresh, cooked alone or with other greens, or added to soups and stews. You can chop them finely and add them to salads. In mild winters they serve as cold-season greens. Add fresh flowers to salads and strew them as garnish on soups or cold drinks. They look great in pickles, aspic, gelatin desserts, and even cake frosting. You can candy them in egg whites and sugar.

**Landscape Uses:** You may want to save borage for marginal areas of the garden. Several other sturdy, invasive plants—mint, lemon balm, and day lily—also grow well in partial shade. You can plant them together where other things will not easily grow.

You should place borage in back of other plants, so that its flowers are visible and the lower leaves and stems are hidden. It goes nicely with plants that are loosely shaped, but seems wild and messy in contrast to plants of a precise growth habit, such as curly parsley. Bor-

3-14. Burnet.
(From Mm. Vilmorin-Andrieux's *The Vegetable Garden.*)

age's flowers are beautiful with bright greens of parsnip, carrot, and plain-leaf parsley. Forget-me-nots, cornflowers, and light blue petunias, such as 'Mercury,' can be planted nearby.
**Varieties:** Rare borage varieties have variegated leaves and purple or white flowers.

### **Burnet (*Poterium sanguisorba*)
hardy perennial

**Ornamental Characteristics:** Burnet plants have small, toothed, blue-green leaflets that last through most winters. They form dense, delicate-textured rosettes, 1½ feet tall and up to 3 feet across (45 by 90 cm). They flower for a month in early summer; dull red flowers in hairy-looking balls are borne at the end of naked stems. The flowers are unusual, but burnet's evergreen foliage and growth habit are its best features.
**Ornamental Drawbacks:** Flowers become brown seed balls. They should be trimmed off, along with their bare stems, when plants are finished blooming.
**Adaptability:** Burnet grows easily from seeds sown outdoors in early spring, and blooms late the first summer. It does best in full sun and dry soil, but tolerates partial shade and moisture. It grows where a lot of other things cannot and reseeds itself prolifically. You may have to pull out some seedlings so the burnet bed does not get too crowded. Plants are hard to transplant or divide.
**Insects and Diseases:** Older plants sometimes rot at the crown during wet winters.
**Food Uses:** The leaves taste very much like cucumbers; young ones are best. Add them to drinks, salads, cottage and cream cheeses,

3-15. Cabbage.

soups, cooked greens, and vinegars. It is possible to freeze or dry them, but there is not much need to do so since you can harvest them far into fall and throughout mild winters.
**Landscape Uses:** Burnet is a useful ground cover for partially shaded areas of the garden. It also thrives on land that is sunny and hard to water. Cornflowers and burnet can flourish and multiply together on a hot, dry bank.

You can grow burnet in a bed, or in a neat row as an edging plant. Its frilly rosettes combine nicely with many other plants, although they can get lost near big, bold-textured ones.

The blue-green colors of fennel and dwarf kale, and the bright greens of carrot, parsley, and celery all go well with burnet. Deep rose, pink, and white flowers are attractive nearby; dianthus is excellent.
**Species:** Great burnet *(P. officinale)* has similar food uses. Its leaves are similar to those of burnet, but it grows taller and has purple flower heads.

## **Cabbage (*Brassica oleracea,* Capitata Group)
hardy biennial grown as an annual

**Ornamental Characteristics:** At its best, cabbage is one of the finest ornamental food plants. Its colors are beautiful and various: gray-green or blue-green, sometimes with crisp white veining, or blue-gray with rose-purple veining and a purple central head. Outer "wrapper" leaves may be plain-textured with wavy edges, or, in savoy cabbages, heavily blistered and bumpy to an extent that is truly baroque. The central heads of cabbages can be spherical, rather broad and flat, or pointed at the top ('Jersey Wakefield'). Wrapper leaves frame the cen-

**3-16.** Cabbage at its worst.

tral head in overlapping layers, like the outer petals of an opening rosebud. Although cabbages are more hefty than any imaginable rosebud, 1 to 2 feet (30 to 60 cm) tall and 2 to 3 feet (60 to 90 cm) wide, their rosette shapes can be perfectly symmetrical and lovely.

**Ornamental Drawbacks:** At its worst, cabbage is quite ugly. Its colors and leaf textures are consistently attractive, but its growth habit is something else. (See figure 3-16.) The rosette shape is so definite that a row or mass planting looks bad if plants are not fairly uniform, and few plants are capable of being as nonuniform as cabbage. Lack of uniformity can be caused by uneven thinning or uneven environmental conditions. It may occur if the cabbages do not grow at a fast, steady pace. Slow growth may be a result of unfavorable gardening practices, insects and diseases, or climate.

Some varieties never acquire an attractive shape, usually because they have sparse wrapper leaves, a characteristic of some dwarf and compact cabbages. Insects and diseases are the biggest drawbacks to using cabbages as ornamentals.

**Adaptability:** Cabbages can be planted in early spring or late summer in many parts of the country. It is important to plant at the correct time for your climate and to use varieties that will be ready at the suggested local harvest time. Cabbages are strongly influenced by temperature, and excessive cold when they are small can cause bolting. Good growing conditions are especially important for cabbages, because they do so much better when they grow rapidly and evenly.

After they are a good size, they can weather the cold down to about 20°F. (-7°C.).

**Insects and Diseases:** Many cabbage problems are widespread, and it is very likely that at least one or two of them will show up in your garden. Many diseases and insects affect the plants in ways that are either fatal or permanently disfiguring. Flea beetles, cutworms, cabbage maggots, aphids, cabbage loopers, cabbage worms, and harlequin bugs are possible insect problems. Since cabbage wrapper leaves grow in layers that overlap only partially, insect-chewed leaves near the bottom of a plant remain clearly visible until harvest. Constant vigilance may keep damage to a minimum. I tried surrounding cabbages with companion herbs, such as wormwood, sage, and thyme. Looked great. Did not keep away cabbage butterflies or other insects.

The most serious diseases are club root, yellows, black rot, black leg, and nematodes. Some cabbage varieties are resistant to yellows.

**Food Uses:** You can store cabbages in cool basements or leave them in the garden during mild falls and winters. Eaten raw in salads and coleslaws, cabbages are an excellent source of vitamin C. They can also be cooked fresh, canned, or pickled as sauerkraut or spicy Korean kim chee. Use wrapper leaves for stuffed cabbage.

**Landscape Uses:** Cabbage's large, definite shape can make combining it with other plants difficult. Its colors, however, blend well with many other colors—blue, blue-green, gray-green, rose, deep red, purple, pink, white, and yellow.

Cabbage looks great in beds or formal rows, *if* it is fairly uniform. Plants used nearby should be quite big—cleome, blue salvia, tall zinnias. Kale and fennel are effective contrasts, because they have large-scale but delicate foliage. Chrysanthemums and cabbage are excellent fall companions, and spring cabbages look very attractive with spring-flowering bulbs.

In fall, cabbage heads can be left for a while and harvested one at a time; but in spring they often have to be harvested fast before they split. A group of cabbages disappearing from your garden all at once can leave a big visual gap.

**Varieties:** Green cabbages with good coloring and uniformity include 'Stonehead,' 'Express,' and 'Early Jersey Wakefield.' Japanese 'Mars 77' is exceptionally good-looking, with very dense, uniform growth, and pale blue-green leaves that have a strong network of whitish veins. 'Defender' and 'Tastie' are especially unattractive varieties.

'Red Danish,' 'Mammoth Red Rock,' and 'Red Head' are red cabbages with vivid coloring, good uniformity, and attractive growth habit. They mature in eighty-five to ninety-two days (compared to sixty-seven days for 'Stonehead').

Outstanding savoy cabbages (see figure 3-17) are European

**3-17.** Savoy cabbage. (From Mm. Vilmorin-Andrieux's *The Vegetable Garden.*)

'Yslanda' and 'Ice Queen.' 'Ice Queen' is available in U.S. catalogs. Both varieties have uniform growth, densely blistered leaves, and a deep blue-gray color. 'Yslanda' is an unusual steely blue. 'Savoy King' is also attractive. Savoys take from seventy-five to eighty-five days to mature.

## *Cabbage, Chinese (*Brassica rapa,* Chinensis Group and *B. rapa,* Pekinensis Group)
hardy annuals

**Ornamental Characteristics:** Chinese cabbages belong to two distinct types. *B. rapa,* Pekinensis Group, is the most common in United States seed catalogs; it is listed as Chinese cabbage, Chinese celery cabbage, wong bok, or pe tsai. It has pale green leaves, and broad white leaf stalks merging into them. It forms solid, barrel-shaped heads, sometimes short and squat. To me it looks heavy and lumplike and its colors are anemic; it is not much of an ornamental. 'Michili' looks slightly better than other varieties (See figure 3-18), if you want to grow it for food. It stores well in cool conditions and has a mild flavor and crisp texture delicious in stir-fries.

*B. rapa,* Chinensis Group, is variously known as bok choy, pak

**3-18.** Wong bok 'Michili.'

3-19. Bok choy. (From Mm. Vilmorin-Andrieux's *The Vegetable Garden.*)

choy, Chinese cabbage, and Chinese mustard cabbage. (See figure 3-19.) Plants are loose, erect clusters of glossy, dark-green leaves with long, narrow, white stalks. They are up to 14 inches tall and 20 inches wide (35 by 50 cm). Bok choy has an attractive, vase-shaped growth habit and good contrast between its bright, shiny leaves and pale stalks. Central stalks are erect and outer ones arch gracefully. If you harvest the drooping outer stalks, the plant is uniformly erect. Attractive, bright yellow flowers may form in the center of the plant.

**Ornamental Drawbacks:** Some cabbage family insects and diseases can disfigure bok choy.

**Adaptability:** Bok choy usually is planted in fall. Spring plantings are sometimes unsuccessful, because some varieties bolt in response to long, cool days when they are young. Bok choy grows best if it grows fast. It can withstand light frost.

**Insects and Diseases:** Chinese cabbages in general seem less hard-hit by insects and diseases that attack other members of the cabbage family. Flea beetles, aphids, and cabbage worms may still be problems, however.

**Food Uses:** Bok choy tastes mild, crisp, and juicy, and is chewier and stronger-tasting than wong bok. Bok choy stores well for only a short time. You can eat it fresh or dry it for later soaking and cooking. In stir-fries, bok choy stalks are added first for longer cooking and leaves are thrown in at the last minute.

You can eat the outside stems and leaves of the plants while the middle keeps growing. When the first flowers appear, the leaves still taste good and the flowers are edible. Several varieties are grown specifically for their flowers.

**Landscape Uses:** Grow bok choy where the whole plant is visible.

3-20. Calendula.

White flowers of baby's breath nearby accentuate the color of bok choy stalks. Yellow flowers, or shiny green leaves such as lettuce's, complement its leaves. Midsummer calendula plantings will bloom in fall when bok choy is in its prime.

**Varieties:** 'Chinese,' 'Japanese White Celery Mustard,' 'Japanese Giant White Celery Mustard,' and 'Crispy Choy Loose-leaved Chinese Cabbage' are very similar in appearance. 'Crispy Choy' and 'Chinese' are more stubby-looking, because their stalks are shorter and thicker.

There are several varieties of flowering bok choy, or hon tsai tai. The most attractive, purple hon tsai tai, has dull green leaves with purple midribs, and small, bright yellow flowers on tall, erect stems. The plant's growth is rather sparse, but colors are subtle and pleasant.

## **Calendula/Pot Marigold (*Calendula officinalis*)
hardy annual

**Ornamental Characteristics:** Calendulas have plenty of bright daisylike flowers, usually orange or yellow. They bloom for a long time in cool weather.



**Ornamental Drawbacks:** Calendula leaves are an unremarkable grayish-green, long, narrow, sticky, and susceptible to powdery mil-

dew. The flowers quickly become wilted and must be constantly removed to keep plants looking good and blooming heavily. Many varieties have a loose, weedy growth habit.
**Adaptability:** Calendulas can be started early indoors, or sown outdoors as soon as the ground can be worked. They thrive in cool weather and can take some frost. They often reseed themselves; flowers that new plants produce are often single.
**Insects and Diseases:** Large beetles sometimes spend the summer sitting on the flowers. They don't hurt the flowers, but they are not decorative. Other possible problems are aphids, white fly, spider mite, powdery mildew, stem rot, leaf spot, and cucumber mosaic virus brought by aphids.
**Food Uses:** Calendulas were known as marigolds or pot marigolds in the Middle Ages, when their petals were a popular seasoning. They taste stronger when dried, and add flavor and golden color to soups, stews, poultry dishes, and rice. They are an inexpensive substitute for saffron.
**Landscape Uses:** Calendulas bloom quickly from seed in cool weather and provide cheerful color to go with the foliage of spring and fall food plants. If you plant masses of calendulas, you may exhaust yourself trying to trim off wilted flowers. You can plant them in back of something that will later grow tall and hide their messiness. They will go to seed and bloom again next spring. Orange calendulas make a strong combination with blue-green kale, or bright green mustard, mizuna, or lettuce. Yellow calendulas are effective in a milder way.
**Varieties:** Some dwarf calendulas have neat, dense growth. 'Orange Coronet' is an excellent variety up to 1½ feet tall and 2½ feet wide (45 by 75 cm), with a crisp-looking, erect growth habit, and rounded, fully double flowers. Unfortunately, it seems very susceptible to powdery mildew in late summer. 'Lemon Coronet' has yellow flowers. There are new dwarf calendulas—'Dwarf Gem,' 'Juliette,' and others—that are worth trial.

'Pacific Beauty' has a loose, weedy habit; its flowers look ragged because the petal edges are unevenly toothed. It is less susceptible to powdery mildew, but in late summer its branches collapse.

Varieties labeled 'Orange King' and *C. officinalis* look very similar. The flowers are somewhat double and are a dark, attractive shade of orange that grows deeper toward the outside edge of the petals. Growth habit is similar to that of 'Pacific Beauty.'

Many other calendula varieties deserve trial. Some have dark centers, and cream-colored or pale orange flowers; petals are sometimes curled. Single-flowered calendulas, preferred by herbalists and cooks, are most often seen not in seed catalogs, but resowing themselves in home gardens.

3-21. Carnation.

## **Carnation/Clove Pink *(Dianthus caryophyllus)*
hardy perennial often grown as an annual

**Ornamental Characteristics:** Carnation's narrow blue-green leaves are quite attractive and give nice color to mild-winter gardens. The beautiful, fragrant flowers bloom for several months in cool summers. They may be single or double, large or small, white, pink, red, purple, even orange and yellow, and variegated combinations of these colors. Some flowers have petals with slightly toothed edges; others have petal edges that are more deeply cut in a rather graceful fringe.
**Ornamental Drawbacks:** Long-stemmed carnations may need staking.
**Adaptability:** Carnations can be propagated from seed, cuttings, or division. In most climates they must be started indoors, since they take about four months from seed to bloom. They are well adapted to cool summers; blooming slows in hot weather. Plants do not survive outdoor winters colder than about 10°F. (-12°C.).
**Insects and Diseases:** Insect and disease possibilities include spider mites, thrips, cabbage loopers, and cutworms; several wilts, stem rots, rusts, and leaf spots. Many carnations are nevertheless very healthy.
**Food Uses:** Carnation petals, with the bitter white inner tips cut off, lend mild clovelike flavor to a variety of foods. Carnation syrup and conserve are two possible uses; in his herbal John Gerard (see Bibliog-

raphy) recommends the conserve because it "wonderfully above measure doth comfort the heart." Leaves may be poisonous.

**Landscape Uses:** Neat bush carnations are great edging plants, especially in combination with other plants that are showy early in the season and can recede into the background or be pulled out later. Dill 'Bouquet' with its finely cut, blue-green foliage is an excellent partner of this sort and so is lavender. Carnations can be planted near other flowers of similar colors and near many other *Dianthus* species. Long-stemmed carnations are most dependable in the background, preferably near sturdy plants they can lean on. They make an attractive combination with blue-green or purple cabbage.

**Varieties:** Carnation (*D. caryophyllus*) is the *Dianthus* species traditionally used as an herb. 'Juliet' is a compact bush, about 1½ feet tall and 2 feet wide (45 by 60 cm), with fragrant red flowers that are fully double and 2 inches (5 cm) across. 'Fragrance' is also compact and sweet-scented, with slightly smaller flowers of pink, red, and purple. 'Chabaud's Giant Improved' blooms a week or two later than the others. Very much like flowershop carnations, it has large, double, fringed flowers in a beautiful range of colors from white and apricot to deep red. Plants grow up to 2½ feet tall and 3 feet wide (75 by 90 cm). Their flower stems lean but never wholly collapse. The plants look disheveled and rather fragile, with long, wandering stems, and narrow, sometimes curling foliage.

Although 'Chabaud' has the strongest scent, the varieties all taste the same, slightly bitter and clovelike.

**Species and Varieties:** Many other varieties of *D. caryophyllus* deserve trial, as do hybrids between *D. caryophyllus* and other species. Single-flowered forms are preferred by herbalists.

Several sturdier *Dianthus* species are as fragrant as *D. caryophyllus* and may make equally good culinary herbs. One herbal suggests *D. plumarius* in particular. Some *Dianthus* bloom well in hot summers and survive cold winters. Some are evergreen and some are excellent ground covers.

## ***Carrot (*Daucus carota* var. *sativus*)
hardy biennial grown as an annual

**Ornamental Characteristics:** Carrots have dark bright green leaves, finely divided into a delicate fernlike texture. Many varieties have strong, bushy tops up to 1½ feet tall and 2 feet wide (45 by 60 cm); they are erect early in the season and arch over at the edges late in summer. Carrots can stay in the garden through fall and mild winters, providing cheerful foliage color until hard frosts. If they are not harvested, lacy, white flower umbels will bloom the next spring.

3-22. Carrot.

**Ornamental Drawbacks:** Some varieties, especially small-rooted ones, have sparse, weak tops.
**Adaptability:** Carrots can be sown outdoors from early spring onward, although low temperatures when plants are small may increase bolting. They can be sown thickly for later thinning. They germinate slowly and need loose, moist soil in which to do it. Use short, thick-rooted varieties for heavy or shallow soil. If you harvest carrots promptly when (or before) they are full-sized, you will have no complaints. If you leave them in the ground after they mature, they may crack and/or rot, especially in wet, warm soil. Carrots tolerate some shade, but the tops may be spindly.
**Insects and Diseases:** Most carrots are very healthy, with few serious insects or diseases. Possible problems are carrot rust flies, carrot weevils, wireworms, blight, and yellows. On the other hand, the splendid caterpillars of swallowtail butterflies are delightful visitors—green and plump with black stripes and yellow spots—and seldom eat enough foliage to damage carrots.



**Food Uses:** The uses of carrots are well known. You can store them for long periods in cool, moist cellars or in the ground. Variety differences in flavor increase as carrots are stored in the ground. Some

carrots have little flavor, some taste slightly bitter, while others are sweet and mild. A few varieties have a sweet, strong, soapy taste.
**Landscape Uses:** A vigorous carrot variety can be grown with almost anything. Some of the best companions are Queen Anne's Lace, blue lace flower, linaria, 'Lulu' marigold, *Cosmos sulphureus* ('Diablo,' 'Sunset,' 'Bright Lights'), coriander, and parsley. Other good combinations are various ferns, fennel, parsnip, celery, and pepper. The only plants to avoid are large plants with coarse foliage. Carrots can be planted in a feathery mass or grown as an edging, alone or with other plants.
**Varieties:** The most vigorous erect varieties are 'Autumn King,' 'Danvers Half Long,' 'Royal Chantenay,' 'Scarlet Nantes Strong Top,' 'Spartan Bonus,' and 'Strong Top Nantes.' 'Autumn King' has slightly broader, bolder leaflets that give it a dense luxuriant texture. Most of these varieties hold for a while when stored in the ground, but 'Strong Top Nantes' cracks easily and should not be used for this purpose. (The champion storing-in-the-ground variety is 'Target'; looks and flavor are acceptable, but not top quality.)

Good-looking carrots that taste best—sweet and fairly mild—are 'Autumn King,' 'Scarlet Nantes Strong Top,' and broad-rooted 'Royal Chantenay.' Of these three, 'Royal Chantenay' is most resistant to cracking.

Short-rooted carrots have a definite place in container plantings and in very heavy soil. However, the worst-looking carrots in our trial gardens were short-rooted 'Bunny Bite' and 'Short 'n' Sweet.' Their poor growth may have been at least partially a result of susceptibility to motley dwarf virus, a disease of the western United States that dwarfs tops and roots of carrots.

## **Celery *(Apium graveolens* var. *dulce)* and *Celeriac/Celery Root *(A. graveolens* var. *rapaceum)* hardy biennials usually grown as annuals

**Ornamental Characteristics:** Celery and celeriac have beautiful, glossy, bright green leaves, divided into leaflets with evenly toothed edges that give the plants a crisp texture. Some celery varieties grow to be dense, erect masses about 15 inches (38 cm) tall.
**Ornamental Drawbacks:** Other celery and celeriac varieties have sparse foliage, often combined with spreading stalks. Plants look somewhat meager and awkward. Many of the diseases and mineral deficiencies to which celery and celeriac are susceptible disfigure the leaves.
**Adaptability:** Mineral deficiencies result in cracked stalks, chlorosis, or black heart. There is varietal resistance to some disorders.

3-23. Celery—'French Dinant' type.
(From Mm. Vilmorin-Andrieux's *The Vegetable Garden.*)

Celery is difficult to grow. It germinates and grows slowly; it should be started indoors or in a hot bed or cold frame, under cool, moist conditions. (Not *too* cool, however; ten days' exposure to temperatures of 40° to 55° F. [4° to 13° C.] can cause bolting in susceptible varieties.) Celery is a native of marshy places and needs a good deal of moisture and nutrients. It requires a long, cool growing season. Blanching the stems is not necessary; varieties are eaten green-stemmed or they are "self-blanching"—pale even under ordinary growing conditions.

Celery and celeriac live through fall and mild winters, but the stalks become stringy and may be hollowed out by slugs. Overwintered celery produces broad, greenish-yellow flower umbels on 2-foot (60-cm) stalks in May. The flowers are prominent but not especially attractive. Seeds ripen in midsummer.

**Insects and Diseases:** Plants may get several blights, leaf spot, root rot, brown rot, fusarium yellows, and virus diseases transmitted by aphids and other insects. Varietal resistance is available for most of these diseases. Destructive insects are carrot rust fly and tarnished plant bug.

**Food Uses:** Celery stalks are frequently bitter, stringy, or pithy. Good-tasting stalks can be canned or used fresh. They can be stored for long periods in cool, moist conditions. Use leaves, seeds, and poor-quality stalks fresh or dried, as seasonings in soups, stews, and casseroles.

Celeriac has an enlarged stem base and root, which is cooked and eaten in stews or salads. It is used in European cooking, and is the "apio" of Latin American markets and cuisine.



**Landscape Uses:** The bright green foliage and erect growth of good celery varieties have many uses in the landscape. Other bright green leaves, such as those of parsley, carrots, and corn, look good with celery, as do blue-greens such as those of kale and fennel. Celery foliage is a good foil for bright flowers and fruits of yellow, orange, red, and purple. You can grow celery near pepper varieties whose fruits turn color early. Celery, 'Bronze Rocket' snapdragon, and green-leaved corn make an outstanding combination.

Fine-textured foliage plants are excellent with celery; so are fine-leaved flowers such as signet marigold, linaria, baby's breath, German chamomile, and matricaria. The delicacy and purplish shades of blooming summer savory are an excellent contrast to celery. In mild climates overwintered celery is a great background for early spring bulbs.

Bushy celery varieties can be grown massed in a bed, with plants at 15-inch (38-cm) distances; they are neat enough for a long-lasting edging. Celeriac and the sparser celery varieties look best spaced closely (about 12 inches or 30 cm) and used as background plants.

**Varieties:** 'Summer Pascal' and 'Utah 52-70R' are excellent celery varieties, vigorous and uniform in growth. 'Utah 52-70R' is slightly better-looking, but 'Summer Pascal' has less susceptibility to bolting and has resistance to several deficiency-related disorders. 'French Dinant' is a seasoning celery with thin stems. It is rather fine-textured and fernlike, unusually dense, vigorous, and winter-hardy; it forms large, fountain-shaped mounds of clear green. It grows up to 2½ feet tall and almost 4 feet wide (75 by 120 cm). Plants similar to it may be listed in seed catalogs as French or Chinese celery. Celeriac varieties are less vigorous than celery; 'Alabaster' looks better than 'Marble Ball.'

## **Chives *(Allium schoenoprasum)*
hardy perennial

**Ornamental Characteristics:** Chives form dense clusters up to 1½ feet tall and 2 feet wide (45 by 60 cm). The deep green leaves are long, thin, and spiky-looking. They die back each winter and in early spring new spears poke through the soil. In May the plants are covered with pretty lavender flower balls.

**Ornamental Drawbacks:** Central leaves are erect, but older ones on the outside may fall over, giving the plants a floppy, rather messy appearance. This is less likely to happen when plants are grown in full sun.

**Adaptability:** Chives are slow to start from seed, but you can easily buy new clumps or divide old ones in early spring. In partial shade, plants may look scrawny and have fewer flowers.

**Insects and Diseases:** I know of none.

**Food Uses:** Both flowers and leaves can be used, fresh or frozen, in

3-24. Chives.

salads and seasonings. The leaves should be cut at ground level from edges and inconspicuous parts of the plant, so it does not look chopped up.

**Landscape Uses:** Combine chives with plants of similar textures—bunching onion, Chinese chives, ornamental allium, and sea thrift. Contrast them with thyme and curly parsley. They are best grown in a bed, where leaning leaves are less noticeable. They are most attractive in spring, when they should be given a prominent place in the garden. Later in summer you can leave them in the foreground if you feel like keeping them trimmed and tidy, or you can let something grow up in front of them.

**Varieties:** Fine-leaved, medium-leaved, and large-leaved forms are occasionally available.

## ***Chives, Chinese/Garlic Chives (*Allium tuberosum, A. ramosum,* and *A. chinense; A. odorum* may refer to *A. tuberosum* or *A. ramosum*)

hardy perennials

**Ornamental Characteristics:** Chinese chives have broader leaves than regular chives, larger flowers, and a neater growth habit. They are tidy plants up to 1½ feet tall and 1½ feet wide (45 by 45 cm). Leaves are ½ inch (1.3 cm) wide, flat, and medium green. They grow erect at the center of the plants and arch over at the edges. (See figure 3-25.)



Chinese chives bloom the first year from seed. The flower heads are round clusters of small, star-shaped flowers, lavender or white.

3-25. Chinese chives.

Some are sweetly scented. Plants spread easily by rhizomes and self-seeding.

**Ornamental Drawbacks:** Chinese chives planted from seed grow very slowly and are scarcely noticeable until they begin blooming in late summer.

**Adaptability:** They can be started from division or seed planted early in spring to ensure good germination and a respectable amount of growth during the first season. Rich soil is best, but plants are flexible in their requirements. They should be divided when crowded. Leaves die back in fall, but roots and small bulbs overwinter easily in most climates.

**Insects and Diseases:** Chinese chives are very healthy.

**Food Uses:** The leaves, flowers, flower stems, seedheads, and small bulbs of Chinese chives are all edible. Use leaves fresh or frozen; use flowers fresh, frozen, or dried. Chinese chives taste like mild garlic and you can use them as you would garlic.

**Landscape Uses:** Mass plants in a bed or line them up as a small-scale edging. They are attractive in pots and planters and combine well with chives, bunching onions, and ornamental alliums. White-flowered forms are attractive with other white flowers, such as coriander, baby's breath, white nicotiana, tuberose, and white morning glory. Rounded, bushy plants, such as curly parsley, provide effective contrast to the leaves. Chinese chives can share space with small spring-flowering bulbs; their leaves are new and small when the bulbs are blooming.

**Species and Varieties:** Two Japanese varieties of *A. tuberosum* are 'Chinese Leek Broadleaf' and 'Chinese Flowering Leek Tenderpole.' 'Broadleaf' is bushier and more vigorous; it blooms earlier. 'Tender-

3-26. Chrysanthemum greens.

pole,' relatively wispy in its growth habit, blooms several weeks later than 'Broadleaf' and continues for a long time.

Most American catalogs offer only seeds of what they call "Chinese Chives," which turns out to be *A. tuberosum.* Its late-summer flowers are white with a brown or green stripe down the center of each petal. Fragrant-flowered garlic (*A. ramosum*) is very similar to *A. tuberosum. A. ramosum* blooms in early summer and has white flowers with a red stripe down the center of each petal. One or both of these species have varieties with variegated leaves. A third species, rakkyo (*A. chinense*), has late-summer flowers that are purple and slightly nodding, and is dormant in early summer.

## *Chrysanthemum Greens/Shungiku
## *(Chrysanthemum coronarium)*
hardy annual

**Ornamental Characteristics:** Shungiku is grown for its leaves, used as potherbs in Oriental cooking. The leaves are gray-green and finely divided. After several months the first daisylike flowers appear and soon cover the plants. They are buttery yellow, single, and about 2 inches (5 cm) across. Sometimes the petals are fringed with white along the outer edges. Plants flower profusely for six weeks or more.

They are most attractive when old flowers are cut off, but this is not necessary. They bloom for a long time even without trimming, and wilted flowers become tan seed balls which do not detract much from the plants' appearance. When they have flowered, the plants die; new seedlings sprout around them to bloom in fall.

**Ornamental Drawbacks:** Shungiku should be eaten before it flowers, although, at that time, it is not yet very attractive. Once flowering starts, it is showy, but its main use as a food plant is over. Sow shungiku thickly, eat thinnings, and leave a few plants to decorate the garden and reseed themselves.

Flowering plants shoot up to a height of 2½ feet (75 cm) and become top-heavy with blooms; they may need to be staked. Some of the flowers have ray florets of uneven lengths. Individual blossoms may have a ragged appearance, but they still look good in a mass.

**Adaptability:** Shungiku tolerates light frosts. It grows and tastes best in cool weather. Early spring sowing leaves time for it to reseed itself and produce a second crop. It can be grown in partial shade as a leaf crop, but full sun is best if you intend to leave it to flower. Its growth habit will be as erect and compact as possible.

**Insects and Diseases:** Shungiku may sometimes be prey to diseases of ornamental chrysanthemums—black aphid, gall midge, wilt, rust, stunt virus, and powdery mildew—but it usually is quite healthy.

**Food Uses:** Young shungiku leaves are used in Japanese soups and sukiyaki and in Chinese stir-fries. They taste fragrant and slightly astringent.

Shungiku flower petals are edible, but are more bitter than Oriental chrysanthemum varieties developed for good-tasting flowers. Use them fresh, pickled, or dried, for soups and stir-fries.

**Landscape Uses:** Shungiku is a good background plant. Its leaves have no need to be showy, and in midsummer its cheerful flowers can rise above foreground plants such as small marigolds and calliopsis, German chamomile, and yellow-fruited peppers. Dwarf calendula and shungiku can bloom together in fall, about two months from seed. Layia, orange cosmos, and coreopsis have flowers similar to those of shungiku. Blue salvia and small purple asters provide good fall contrast.

**Species and Varieties:** There are large- and small-leaved forms of shungiku. Ornamental varieties of *C. coronarium*—'Golden Gem,' 'Tom Thumb', and others—can be tried for greens.

Many chrysanthemums grown in Japan for good-tasting flowers ('Niigita' and others) are not yet available in the United States but they and ornamental chrysanthemums are derived from *C.* x *morifolium.* If you taste ornamental varieties in search of good-flavored flowers, start with yellow ones, traditionally preferred for culinary use in the Orient.

3-27. Coriander.

### ****Coriander/Cilantro** *(Coriandrum sativum)*
hardy annual

**Ornamental Characteristics:** Coriander grows up to 15 inches tall and 15 inches wide (38 by 38 cm) in erect bushes with many branches. Its dark green leaves are divided into leaflets. Older leaflets at the base of the plant are almost round, with deeply toothed edges; the young, upper leaflets are feathery and narrow. Coriander flowers are lacy compound umbels which appear to be lavender; actually, each of the small umbels has white flowers on its edges and purple ones in its center. Coriander blooms for two to three weeks; its flowers are followed by round, shiny green seeds. During the next few weeks, the foliage turns red and purple, seeds become brown and ripe, and plants die.
**Ornamental Drawbacks:** Despite its beauty, coriander has a fast life cycle—two and one-half months from seed to death of the plant. Some coriander plants fall over in full bloom. Close planting—6 inches (15 cm) or less between plants—will discourage this.
**Adaptability:** Coriander does not transplant well. In early spring it should be sown thickly where it is to grow; you can eat the thinnings. For a constant supply of leaves, make several plantings. Full sun and pinching out the central growing tips encourage bushiness. The plants survive light frosts.
**Insects and Diseases:** In the western United States, coriander

sometimes gets motley dwarf virus, which dwarfs the plant and turns it yellow and purple.
**Food Uses:** The leaves are popular in Chinese, Southeast Asian, Indian, Italian, Caribbean, South American, and Mexican cooking. They are used in salads, guacamole, chutneys, soups, stews, stir-fries, and rice. They taste best when the plant is small and you can use them fresh, dried, or frozen. They smell pungent; their flavor is not far from parsley's, but unique. Not everyone likes it.

In Scandinavian baking, coriander seeds, which smell and taste spicy, sweet, and citruslike, are used whole or ground. They are also used in spice mixes for pickles, sausages, and Indian curries.
**Landscape Uses:** Coriander is best grown in a closely planted bed near the foreground of a border. But if you plant it in the spring, you may want it to be unobtrusive later in summer as its seeds ripen. One way of arranging this is to plant a narrow, easily hidden swath of coriander early. Several weeks later, sow other, taller plants in front. By midsummer the taller plants will hide most of the coriander. Carrots or plain-leaf parsley are good for this purpose.

Plants with large, bold leaves or flowers overwhelm coriander, but many closer contrasts, such as thyme and savory with their small, plain leaves, and linaria and dwarf annual phlox with small flowers and neat foliage, are excellent. Feathery-textured plants such as baby's breath, didiscus, and copper fennel also look good.
**Varieties:** There are no varieties available in United States catalogs.

## ***Corn *(Zea mays)*
tender annual

**Ornamental Characteristics:** Corn is a boldly effective plant. It is usually quite large—4 to 9 feet (1.2 to 2.7 m) tall and up to 4 feet (1.2 m) wide. Its leaves are big, plain, and strong-colored—reddish brown, or grass green with white midveins. Leaves grow up from the main stalk and gracefully arch over so the plant is shaped like a fountain. Erect, branched tassels, the male flowers, rise above the leaves. (See figure 3-28). Stringy silks, the female flowers, which droop from the end of the husks, are less conspicuous.
**Ornamental Drawbacks:** Several varieties, notably 'Queen Anne,' develop abortive ears on some tassels. These look peculiar, but can be cut off. 'Golden Midget' sweet corn and 'Fireside' popcorn are dwarf varieties so diminutive and sparsely leaved that they look deformed.

Corn plants grow bigger when they get plenty of water. A variety grown in one location may be several feet shorter than the same variety grown in another location with higher soil moisture. This can be a problem if you are trying to grow a uniform hedge.
**Adaptability:** Corn tolerates a wide range of environmental condi-

3-28. Corn.

tions. There are sweet corn varieties that will mature ears from seed planted outdoors after danger of frost in all parts of the United States except Alaska. Popcorn and field corn take longer. Seed germination may be difficult in wet or cold soil, and animals sometimes eat the seeds.

Corn is wind-pollinated; for best ear development, it should be planted in blocks or double rows. Pollen from neighboring varieties can affect kernel quality. Popcorn and field corn should be separated from sweet corn by 100 feet (30 m) or four weeks' difference in planting time, so that sweet corn is not made starchy by popcorn or field corn pollen. "Extra-sweet" corn varieties should also be isolated from standard sweet corn.

Corn has flexible requirements in terms of plant spacing. Plants grown 3 feet (90 cm) apart in each direction branch and produce more ears per plant. Plants grown 1 foot (30 cm) apart branch less and produce fewer ears per plant, but more ears per square foot of growing space. Below 1 foot apart, production drops because of barren stalks.

Corn uses heavy amounts of soil nutrients; if you plant it in the same place two years in a row, fertilize generously.

**Insects and Diseases:** Corn usually is quite healthy, but possible insects and diseases include cutworms, earworms, Southern root cornworms, white grubs, corn borers, wireworms, flea beetles, seedling root rot, corn smut, and Stewart's bacterial wilt.

**Food Uses:** Popcorn and field corn are grown until their kernels are dry. Field corn is then fed to livestock or is ground up for cornmeal.

Sweet corn has many well-known uses. You can eat it fresh, frozen, canned, and pickled in relishes. For the space it occupies, corn is not a high-yielding vegetable. Under most conditions, it produces one or two ears per plant.

**Landscape Uses:** Because of its size, corn can be used for temporary large-scale structure in a garden. It is a good backdrop for a mixed border. It can be planted as a summer hedge or screen, along with sunflowers or sorghum. It can create a dramatic tall entrance to a space, or direct a view or the flow of a path. A space in the middle of a mass of corn can be a private summer cave for children.

Make sure other plants that grow next to tall corn are at least 3 feet (90 cm) high, so they are not overwhelmed by the vertical effect of corn. (Also, the lowest part of corn plants is the least attractive part, and this way it is hidden.) Corn should be planted to the north side of shorter plants, or at least 3 feet (90 cm) away from them, so they are not growing in its shade.

Closely planted corn beds, with 1 foot (30 cm) between all plants have a dense, massive look. Leave room enough around the beds to reach in and harvest. Corn also looks very good growing in rows or hills.

The crisp colors of green-leaved corn blend with almost any colors except gray-green. They are cool-looking against a white wall. Light and heat reflected from the wall increase the amount available to the corn. Red-leaved amaranthus, 'Rhubarb' chard, and red-leaved lettuces look good with red-leaved corn. Red-leaved and green-leaved corn can be grown together; the contrasting leaf and tassel colors are beautiful, especially against a plain background—sky, white walls, or evergreen foliage. Plants that echo the shape of corn are celery, fennel, and carrots. Corn's texture blends with other plain, bold textures or with contrasting delicate ones.

**Varieties:** Many sweet corn varieties are available, and most look very attractive. Some look slightly better than others, but the main differences are in harvest dates and in sweetness and tenderness of kernels.

Vigorous, strong-colored varieties, most about 6 feet (1.8 m) tall, include 'Bellringer,' 'Golden Cross Bantam,' 'Iochief,' and 'Stylepak.' They are all dense-growing, with yellow kernels, cream-colored tassels, and bright green leaves with prominent white stripes. 'Burgundy Delight' is green-leaved early in summer, but as its ears develop, it turns mahogany red all over. Its kernels are yellow and white. 'Tendertreat' has reddish-green plant coloring that is not outstanding, but the plant is impressive because of its height—up to 9 feet (2.7m).

None of these varieties is especially high- or low-yielding. The sweetest and most tender are 'Bellringer,' 'Stylepak,' and 'Burgundy Delight.' 'Bellringer' is earliest to yield.

'Sweet Sue,' 'Sugar Loaf,' and 'Silver Queen' are all attractive and good-tasting, but slightly inferior to the others in coloring and vigor. 'Silver Queen' has white kernels, and plant coloring similar to that of 'Burgundy Delight.'

'Burpee's Peppy' and 'Strawberry' popcorns both looked okay in our trial gardens, but matured late. Many varieties of popcorn and field corn are worth trial in areas with long, hot summers.

## **Cucumber (*Cucumis sativus*)
tender annual

**Ornamental Characteristics:** The sharply toothed, maple-shaped leaves of cucumber are crisp and attractive. Its strong vining habit and curling tendrils give it an air of wild vigor. (See figure 3-29.) Vining cucumbers can be grown as ground covers or climbers. When grown on the ground, they are 15 to 24 inches (38 to 60 cm) tall and 3½ to 5 feet (1.1 to 1.5 m) wide, depending on variety. There are several bush varieties that are 8 to 10 inches (20 to 25 cm) tall and 1½ to 2 feet (45 to 60 cm) wide.

Most cucumbers have male and female flowers on the same plant. Many flowers are hidden in the foliage, but some male flowers are clearly visible on top of the plants. They are bright yellow and about 1½ inches (4 cm) across; they have fused, pointed petals.

**Ornamental Drawbacks:** Fruits are inconspicuous, since they are usually green and borne underneath the foliage. Late in summer the leaves are sometimes disfigured by a number of diseases.

**Adaptability:** Cucumber seeds should be planted in warm soil. Their growing season is short—fifty to sixty-five days from seed to first fruit—and in most places it is not necessary to start them early indoors. They grow best in well-watered soil that contains plenty of organic matter. Excessive heat or drought can cause bitter flavor; some varieties have a greater tendency than others to become bitter.

**Insects and Diseases:** Diseases include bacterial wilt, anthracnose, angular leaf spot, downy and powdery mildew, scab, and mosaic. Some varieties are resistant to five or six diseases. Insect pests include striped and spotted cucumber beetles, squash vine borers, squash bugs, aphids, and pickleworms.

**Food Uses:** Plants need to be frequently picked so they will keep producing fruit. You can stuff, marinate, or use the fruit raw in salads and cold soups. It is cooked in Oriental stir-fries and made into many kinds of pickles and relishes.

3-29. Cucumber.

There are two kinds of fruits—slicers and picklers. Picklers stay small; even if they are left on the plants for a long time, they just get thicker and seedier. Slicers are juicy and grow very long and broad if left unpicked. They can be eaten raw or made into pickles.

**Landscape Uses:** You can grow bush cucumbers in containers or massed in beds. Vine cucumbers are excellent ground covers, especially near short-term crops; they quickly spread over into the empty spaces that the short-term crops leave when harvested. Plenty of room should be allowed around cucumber beds so you can harvest fruit without having to walk on the plants.

Vine cucumbers can climb on fences and trellises, alone or with flowering ornamental vines. When grown vertically they take up less room, display their flowers and fruits well, and provide shade, privacy, and screening. They can camouflage unsightly objects such as chain link fences and stumps.

**Species and Varieties:** The best varieties are disease-free and do not have bitter fruits. In our location, the best slicers we tested were 'Pacer,' 'Poinsett,' 'Marketmore 70,' and 'Lemon,' all vining cucumbers that bear late. 'Marketmore 70' at fifty-seven days is earliest of the four. 'Poinsett' has tolerance to five major cucumber diseases. 'Pacer' and 'Marketmore 70' resist scab and mosaic. 'Lemon' is an oval, yellow cucumber extolled in seed catalogs for its "burpless" qualities. 'Bush

3-30. Day lily.

Whopper,' the most attractive bush slicer we grew, had bitter flavor. 'Pot Luck,' another bush slicer, had more leaf disease, but tasted better and bore early.

'Tiny Dill,' a bush cucumber, was the best-looking pickler tested. It tasted bitter, but bitter flavors are masked in pickling. 'Liberty,' a vining pickler, looked fairly good, tasted milder than 'Tiny Dill,' and bore fruit early. It has excellent disease resistance.

West Indian gherkin *(Cucumis anguria)* had poor disease resistance in our location; so did long-fruited Oriental varieties: 'Kyoto 3 Feet' and several others.

## ***Day Lily (*Hemerocallis fulva*)
hardy perennial

**Ornamental Characteristics:** Day lilies are sturdy and adaptable, graced with beautiful flowers, arching leaves, and many edible parts. Plants are at first about 1½ feet (45 cm) wide, and, as their rhizomes spread, they quickly become broad clumps. They are not native American plants, but they have escaped from cultivation and grow wild on old farms and along roadsides in colorful profusion.

Day lily leaves are long, flat blades of medium green. In full sun, they grow erect at the center of the plant and arch over at the edges. In shady conditions, most leaves arch over. In midsummer, plants bear large lily flowers that have broad petals with pointed tips, on 2- to 3-foot (60- to 90-cm) stems. The species has orange flowers, but many new forms have been developed. Flowers are larger, sometimes double, in beautiful shades of red, bronze, apricot, pink, gold, yellow, and cream. Varieties bloom at various times from early summer to early fall. Each flower lasts for one day. Old flowers look wilted, but not especially ugly. There are so many you may as well forget about trying to remove them, for your own peace of mind. Shiny, attractive seed pods soon grow in their place.

**Ornamental Drawbacks:** Some day lily varieties are evergreen or partly evergreen. Leaves may become damaged by winter cold, but stay on the plant anyway, looking untidy until they are cut back or until a new flush of growth occurs in spring. In most climates, varieties that die back to the ground in winter are actually better ornamentals.

**Adaptability:** The amazing day lily can grow in full sun, partial shade, or deep shade. In deep shade it has fewer flowers and its foliage is less erect. It grows best in well-drained soil that is not especially rich. It forms a clinging, spreading ground cover on steep inclines and eroded areas. It is propagated by division in spring or fall.

**Insects and Diseases:** Cutworms, thrips, and slugs sometimes are problems.

**Food Uses:** Flowers, tubers, and tender central stems are used for food. Flower buds and open flowers are used fresh in stir-fries and tempura, or are sautéed as separate vegetables. They can be added raw to salads. Their flavor is mild, pleasant, and slightly mucilaginous. They are good sources of protein and fiber. Buds are dried for Oriental "golden needles" to be used later in soups and stir-fries.

You can boil or steam young stems and eat them as you would asparagus. The small tubers taste crisp, sweet, and slightly smoky. They are especially good in spring, but you can harvest them at any time of year—for baking, steaming, roasting, or stir-fries. They multiply quickly, so there are usually plenty for harvesting.

**Landscape Uses:** Day lilies thrive where many food plants will not succeed. They make a good ground cover under trees, especially summer-flowering ones. They can be planted around shrubs or in a long low-maintenance border with other lower ground covers in front of them. The Chinese plant them at the shaggy edges of cultivated gardens and fields. You can mix different day lily varieties for a range of colors and blooming dates. You can plant them near summer flowers of many shapes and colors.

**Species and Varieties:** Yellow or lemon day lily, *H. lilioasphodelus (H. flava),* has flowers and stems that are useful as food, but roots that are less tuberous than those of *H. fulva.* Leaves are narrower and yellow flowers bloom in late spring. Their fragrance is very much like that of orange blossoms and they have a sweet, flowery flavor. In the Orient, small, yellow-flowered *H. minor,* and *H. aurantiaca,* with orange-brown flowers, are also used for food.

There are over fifteen thousand day lily varieties, and new ones appear every year. Many are of mixed parentage. *H. fulva* and *H. aurantiaca* are the two species used most often in breeding modern day lilies though other species are involved as well. All have edible flowers and young stems, but little is known about which ones have the biggest, best-tasting tubers. A sure bet, however, is *H. fulva* 'Europa,' a variety that grows wild in the northeastern United States. It has vigorously spreading, tuberous roots, and makes an excellent ground cover. Because it is so common, it is seldom seen in catalogs or nurseries. Usually, however, you can find it bursting out of a garden in your neighborhood. Ask the owner for some extra plants. 'Europa' does not have the long and heavy bloom or color range of newer varieties, but its food uses are reliable. If you are interested in eating tubers, stick close to *H. fulva* parentage (if known) and try varieties recommended as vigorous ground covers.

## **Dill (*Anethum graveolens*)
half-hardy annual

**Ornamental Characteristics:** Dill grows up to 2½ feet tall and 1½ feet wide (75 by 45 cm). It has feathery, blue-green leaves and broad, yellow flower umbels. (See figure 3-31) The umbels remain yellow as the flowers wither and seeds begin to develop, and the foliage soon turns yellow as well. The whole plant stays bright-colored for about a month before finally browning. Spring-sown plants are followed by a host of seedlings in the fall and the next spring.
**Ornamental Drawbacks:** It is fortunate that dill stays attractive until the end of its life cycle, because its life cycle is certainly short. This may be partially influenced by environment. Dill seems to live longer and grow bushier before flowering when planted fairly early in spring—in late April as opposed to June. When it flowers fast, it looks wispy and insignificant growing in rows, but it still looks good in beds.
**Adaptability:** Seed germination may be spotty while soil is cool; sow early spring plantings thickly. Dill is hard to transplant and should be sown where it is to grow. It needs full sun to get bushy and does badly when weather is very hot. In windy places it may need to be staked or planted closely. Seed sown every few weeks through the growing season will ensure a constant supply of leaves.
**Insects and Diseases:** Aphids can be a problem and dill is susceptible to motley dwarf virus, a disease of the western United States that dwarfs the plants.
**Food Uses:** Dill leaves ("dill weed"), seeds, and flowers are excellent seasonings, used especially often in Scandinavian countries. All parts are used for seasoning pickles. Leaves often appear in salads, potato and fish dishes, and vinegars, and seeds in sauces and baking.

3-31. Dill.

**Landscape Uses:** Although dill is small and blue-green when young, it will be most visible in the landscape when it is tall and covered with yellow flower umbels. A long-lasting, low planting, such as seed-grown sage, looks good in front of dill. It shows to advantage near blue-green or gray-green leaves and blue flowers; blue salvia is an especially good companion. Dill flowers have a greenish tinge that looks sickly next to intensely yellow flowers or bright green foliage.

Planting dill in thickly seeded beds eliminates the need for staking. Even if early flowering occurs, plants will still look attractive.

**Species and Varieties:** 'Bouquet' and 'Long Island Mammoth' are the two varieties available commercially. Seed labeled simply "dill" may resemble either of them, but most probably 'Long Island Mammoth.' 'Bouquet' has deep blue-green leaves and a relatively bushy growth habit. Once its leaves turn yellow, they stay yellow for a long time before turning brown. Although 'Bouquet' is sometimes sold as "dwarf dill," it grows to the same size or slightly taller than 'Long Island Mammoth.' 'Long Island Mammoth' has greenish leaves and comparatively few of them. It is less vigorous than 'Bouquet' and, in our trial gardens, 'Long Island Mammoth' was affected by motley dwarf virus.

In India and Japan the species *A. sowa* is grown for its leaves and seeds. The seeds are an important curry ingredient. It is taller than dill, with denser flower umbels.

3-32. Eggplant (standard type).

## **Eggplant (*Solanum melongena*)
tender annual

**Ornamental Characteristics:** Eggplants are large, bold-textured, and distinctively colored. They may grow 3 feet tall and 4 feet wide (.9 by 1.2 m), with dense branching and thick foliage.

Most varieties fall into two groups—standard eggplants, which have large, broad-oval fruits and big, gray-green leaves with prominently scalloped edges, and Japanese eggplants, with slender fruits and smaller, plain-edged, medium-green leaves, with deep purple veining. Standard eggplants have lavender stems and flowers; Japanese eggplants have dark purple stems and dark lavender flowers. Both types of eggplants usually have deep purple, shiny fruits borne near the base of the plant and partly hidden by foliage.

**Ornamental Drawbacks:** Plant stems often break in heavy summer rains or as the fruits become heavy. Stems seldom break off completely and the fruit does continue to ripen, but prone branches make the plant structure less attractive. Staking is possible, but may be unsightly unless carefully done. Plants grown in groups help hold each other up.

Other ornamental drawbacks are related to adaptability and disease susceptibility.

**Adaptability:** Eggplants need a long season of sun and warmth to look good and bear well. There are definite varietal differences in both earliness and ability to produce well in cool summers. In cool climates, black plastic increases yield by warming the soil around the plants. The plastic is not attractive, but becomes less conspicuous as the plants spread. In most climates, you must sow early indoors and move them outside in warm weather. If you buy transplants from nurseries, there is seldom much choice of variety.

**Insects and Diseases:** Possible problems include flea beetles, Colorado potato beetles, cucumber beetles, lacebugs, aphids, red spider mites, slugs, fruit rot, and verticilium wilt. Ornamental value is seriously affected by verticilium wilt, a disease likely to develop as the fruit ripens. It seldom affects yield, but leaves of infected plants turn yellow and brown. There are varietal differences in resistance to wilt.
**Food Uses:** Eggplants are a mild-flavored main ingredient of many delicious Italian, Greek, Middle Eastern, Oriental, and Indian dishes. They are eaten fresh or pickled, or can be frozen for later use in casseroles. Green parts of the plant may be poisonous.
**Landscape Uses:** Eggplants grow well along a sunny south wall. They look better massed in a group rather than in a single row or as individual plants. Eggplants have sharp thorns on leaf backs and fruit stems, so leave space around the group so you can easily harvest the fruit. Eggplants can be used as a backdrop with shorter plants in front, although the shiny fruit will be less visible. Because they are susceptible to breakage, they do not work well as a hedge or in a location where vandalism may occur.

If possible, find out which leaf coloring an eggplant variety has—soft gray-green, or green with strong purple veining. Gray-green standard eggplants can be planted near gray-green sage, lavender, bunching onion, and dusty miller. The strong deep reds of amaranthus 'Molten Fire' and Swiss chard 'Rhubarb,' and the purple flowers of nierembergia and heliotrope, are effective with both kinds of coloring. So are purplish red cabbage and blue-green fennel. Fine-textured plants such as cosmos are attractive contrasts to eggplants.
**Varieties:** 'Midnight,' a standard eggplant, avoids the pitfalls of verticilium wilt, breakage, and low yield in cool conditions. Its yield in our trial gardens has been respectable but not heavy, and the plant keeps on looking good until fruits are ripe. Like other eggplants, 'Midnight' branches low and may have to be mulched to prevent fruit rot.

Several other standard varieties with heavier yield and more late-season wilt and breakage are 'Dusky' and 'Jersey King.' 'Black Beauty' did well one year, but quite badly another.

Several Japanese eggplants bear heavily and early even in cool summers (on black plastic). They are beautiful when they begin to bear, but later they become disfigured by breakage and wilt. You can harvest them early when the fruits are small. 'Early Black Egg' is a strong-colored, very erect plant of the Japanese type, but has broad oval fruits. It is especially early and gets bad wilt in late season. 'Small-Fruited Long Tom No. 4' and 'Purple Pickling' have similar characteristics except for slender fruits.



Several dwarf varieties are prone to wilt and breakage. The same is true of varieties with unusual-colored fruits—white, green, and golden.

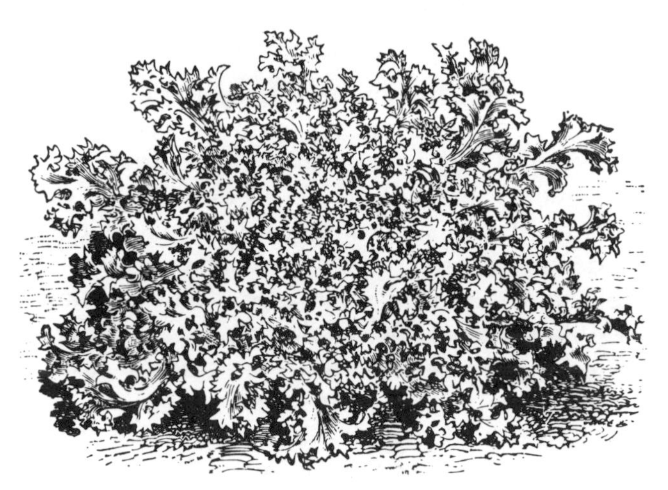

3-33. Endive. (From Mm. Vilmorin-Andrieux's *The Vegetable Garden.*)

## **Endive and Escarole (*Cichorium endivia*)
hardy annuals

**Ornamental Characteristics:** Endive and escarole grow in dense, attractive, rather flat leaf rosettes up to 1 foot tall and 2 feet wide (30 by 60 cm). (Endive and escarole are often confused with chicory [*C. intybus*]. Chicory, included in the list of "Less Attractive Vegetables and Herbs" at the end of this chapter, has long, oval leaves, upright in growth and often jaggedly toothed like those of a dandelion. Various parts of chicory are used for different purposes—salads, coffee, witloof chicory, or "Belgian endive"—depending on which variety is grown.)

Endive leaves are sharply cut and toothed and look very frilly. Leaves of escarole, sometimes called "broad-leaf endive," are broad and rounded, with tiny teeth along the edges. Leaves of endive and escarole are bright light green and so dense that they appear to be jammed tightly together. They resemble some lettuce varieties in their shapes, but last for a longer season than does lettuce. Endive and escarole have attractive blue daisylike flowers borne on erect stalks. By the time they flower, however, their usefulness as food plants has ended.

**Ornamental Drawbacks:** Flower stalks begin to look messy after the first week of bloom. Each flower lasts only one day and soon there are plenty of withered flowers.

**Adaptability:** Endive and escarole last longest and taste best in cool weather, so early spring and late summer are the best planting times. In hot weather they become very bitter and eventually flower, but they are slow (slower than most lettuces) to begin flowering. They are also less susceptible to leaf burn. They withstand frosts down to about 20° F. (-7° C.) In fact, cold weather may decrease their bitterness. Fast growth, aided by plenty of water and soil nutrients, may also decrease bitterness.

**Insects and Diseases:** Slugs and snails sometimes eat the leaves.
**Food Uses:** You can cook endive and escarole leaves or eat them raw in salads. Both taste best when grown in the fall and blanched. This is done by tying the leaves together at the top of the plant or by covering the plants with soil, leaves, boards, flowerpots, or paper. In five to twenty days the plants become pale yellow, more tender, and less bitter.

In French cooking, endive and escarole leaves are braised, or boiled first and then baked with seasonings. Some of the same recipes used for witloof chicory also apply to endive and escarole. Their flavor when cooked is described as nutty and delicious.
**Landscape Uses:** The flat rosettes of endive and escarole are not easy to use in garden design. They make good, neat edging plants, however, and the uniformity of their heads is useful for this purpose. They look good growing in beds, especially near bright green leaves like their own, and near yellow flowers.
**Varieties:** 'Full heart Batavian' is a good escarole. 'Salad King' and 'Green Curled Ruffec' ('Green Curled' or 'Ruffec') are attractive frilly-leaved endives, slow to burn and bolt in hot weather.

## ***Fennel (*Foeniculum vulgare* subspecies *vulgare*)
hardy annual, biennial, or perennial

**Ornamental Characteristics:** Florence fennel and sweet fennel are the most common cultivated forms. Florence fennel, or finocchio, is grown for the thickened bases of its leaf stalks, eaten as a cooked or raw vegetable. Sweet fennel is grown for its large, licorice-flavored seeds. In some countries Florence fennel is called "sweet fennel," which leads to confusion between the two.

Copper fennel has exotic coloring—purplish brown leaves and red-brown stems with pale green bases.

Most fennels are dense, beautiful plants. All have threadlike leaflets like those of dill, but larger and shinier. (See figure 3-34.) Leaflets are usually bright bluish-green, and the sturdy stems are pale aquamarine. Plants are up to 4 feet (1.2 m) wide, and reach 5½ feet (1.7m) in height as their broad yellow flower umbels bloom in late summer. Florence fennel plants are usually smaller than other fennels and they are harvested before they bloom.
**Ornamental Drawbacks:** These plants have none.
**Adaptability:** Full sun is best for fennel, but it can also be grown in partial shade. Plants become sparser, but still thrive and bloom.

Copper fennel germinates well if started early indoors and sometimes blooms its first summer from seed. Florence fennel should be planted in July for a fall crop. If planted in spring, it grows tall, lacks swollen stem bases, and looks identical to sweet fennel. Sweet fennel

3-34. Fennel.

can be planted from early to late spring. All fennels tolerate light frosts. Most fennels are annuals, but some, especially copper fennel, may live as perennials.

**Insects and Diseases:** Aphids, mildew, and slugs are occasional problems.

**Food Uses:** You can use the leaves and seeds of all fennel as seasoning. Seeds of sweet fennel have the strongest licorice flavor. Add them to baked goods, pickles, cheeses, meat, fish, and vegetables. Season soups, salads, potherbs, potatoes, and fish with the leaves. Fennel has many uses in herbal medicine.

Florence fennel's bulbous stem bases, steamed or eaten raw, taste crisp and juicy, with a mild licorice flavor.

**Landscape Uses:** Fennel combines nicely with other feathery-leaved plants, such as cosmos, asparagus, and carrot; with the bold foliage of zucchini squash; with blue-green, gray-green, or bright green leaves; with yellow, orange, purple, blue, rose, or white flowers. It is an excellent background plant or a bountiful hedge. You can grow it in rows, beds, or clumps of several plants together.

Copper fennel looks best with blue flowers and with medium green or brownish leaves. The subtle effect of its coloring can be lost near strong-colored plants.

**Varieties:** Beautiful copper fennel is less vigorous and strongly scented than other fennels. It does not always produce seed. 'Perfection' is a Florence fennel variety, slightly shorter than standard Florence fennel. Carosella is a kind of fennel grown in Italy for its immature flower stalks. "Sicilian fennel" may refer to either Florence fennel

3-35. Flower kale.

or carosella. Wild sweet fennel, naturalized along the West Coast of the United States, resembles sweet fennel, but has smaller seeds.

### ***Flower Kale and Cabbage/Ornamental Kale and Cabbage (*Brassica oleracea,* Acephala Group)
hardy biennials often grown as annuals

**Ornamental Characteristics:** Flower kale and flower cabbage are overlapping terms for one type of plant that is closely related to regular kale. Its variegated leaves range from the plain cabbagelike leaves to frilly leaves like those of dwarf kale. Here the term flower kale describes both types of plants.

Plants grow in neat rosettes up to 1 foot (30 cm) tall and 1½ to 2 feet (45 to 60 cm) wide. Colors are gray-green or blue-green, overlaid with creamy white, pink, purple, or bright rose. The most intense colors are toward the plant centers; bright veining continues outward across the blue-green or gray-green areas of the leaves.

**Ornamental Drawbacks:** Varieties are not always uniform in growth habit. Less uniform types seem very sensitive to differences in spacing. Thin them with great care! Overwintered plants become less attractive as they bloom during spring the second year.

**Adaptability:** Flower kale planted in spring will last until fall in cool-summer regions. In warm climates, sow seed in midsummer so plants avoid bolting but still reach a good size by fall, their best season. As the weather cools, leaf colors become more intense. Plants are undamaged down to about 20° F. (-7° C.).

**Insects and Diseases:** The same insects and diseases that affect cabbages, affect flower kale, but to a lesser degree. If cabbage worms are likely to be a problem, frilly leaves show the damage less than plain ones.
**Food Uses:** There seem to be no differences between flower kale and regular kale in either tenderness or flavor. Both plants are better-tasting in cool weather, especially after frost. Cook flower kale fresh or frozen in the same recipes as kale. Shred it and use it raw in colorful salads.
**Landscape Uses:** Flower kale is striking against a somber background of evergreen bushes or gray walls. Dark pink flowers, white flowers, blue-green and gray-green foliage look good nearby. Aster, chrysanthemum, leek, cabbage, and kale are attractive frost-hardy companions.

A row of flower kale looks strange if it is wildly nonuniform. The right variety, and a double row of it, reduce your chance of this. Mass planting in a bed is another good way to disguise irregularities. You can also grow plants from seed in an inconspicuous place and later transplant the best ones to a prominent place for fall and winter display. They look striking in containers along a porch or path.
**Varieties:** Flower cabbage, flower kale, or both are listed in flower sections of American seed catalogs. Plant colors are often mixed, and sometimes plant forms are as well. In general, seed labeled "flower kale" produces plants with frilly leaves, and seed of "flower cabbage," plants with plainer leaves. It is rare to find a specific variety of one color and form in an American catalog. Many such varieties do exist, leaping in glowing color from the pages of Japanese catalogs, but only two, 'Feather-leaved Coral Queen' and 'Feather-leaved Coral Prince,' are currently available in the United States. Both have deeply cut leaves that form broad, flat-topped rosettes, fairly uniform in size. Their response to crowding is not to stay small, but to orient themselves at an angle on their stems. Those growing on a diagonal look rather like flying saucers. The colors of 'Coral Queen' are rose-purple and blue-gray; 'Coral Prince' is white and blue-gray. 'Coral Prince' had the best cold-hardiness of any variety we tested, suffering only moderate leaf damage at 12° F. (-11° C.).

Harris's strain of flower cabbage, 'Red on Green,' is brightly colored and fairly uniform. 'White Lady' and 'White Christmas' are excellent Japanese varieties not yet available in the United States.

## ***Geranium, Scented (several *Pelargonium* species)



tender perennials

**Ornamental Characteristics:** Scented geraniums have a wide vari-

3-36. Peppermint geranium.

ety of fragrances—rose, lemon, peppermint, and many others. Leaves are usually gray-green in a range of shapes and textures, from shallow-lobed to deeply cut and frilly, and from slightly hairy to soft and plush-textured. (See figure 3-36.) Some scented geraniums have small insignificant flowers. Others have large and showy flowers. Growth habits may be bushy and upright, scrawny and leaning, pendant, or horizontally spreading. Plants grow from 2 to 4 feet (60 to 120 cm) tall. Some are very narrow and others measure 2 to 3 feet (60 to 90 cm) across.

**Ornamental Drawbacks:** Growth habit is important in choosing varieties. Sprawling geraniums and those of stiff, narrow habit are hard to use ornamentally.

**Adaptability:** In most climates scented geraniums need to be replaced every year or to spend winter indoors in a greenhouse or sunny window. As house plants they may receive less light than they need, grow leggy, and become prey to mealy bugs, aphids, white flies, and spider mites. They need a light soil mixture in their pots, and if they are watered too often, their roots rot. Whole plants can be overwintered, or you can take cuttings. However, some kinds take several months to root. Starting them from seed is a very slow process. You may want to save trouble and buy new plants each year.

Full sun and well-drained soil are the best outdoor conditions for scented geraniums. Some kinds, notably peppermint geranium, tolerate partial shade and less than ideal drainage.

**Insects and Diseases:** Scented geraniums are usually healthy under outdoor conditions, though insects that infested them indoors may accompany them outdoors. Disease possibilities are crinkle virus, mosaic virus, bacterial leaf spot, and botrytis blight.

**Food Uses:** Scented geraniums are most often used in sachets and potpourri mixtures, but also lend fragrance and flavor to tea, juice, wine, desserts, soups, sugar, vinegar, jellies, and canned and baked fruits. Both leaves and flowers are good for seasoning.
**Landscape Uses:** For full sensory effect, plant scented geraniums near places where people sit or are likely to brush against as they pass. Raised planters are especially effective. They look good with other lobed or fine-textured foliage and with rose, pink, blue, or white flowers. Fuzzy peppermint geranium leaves and wispy licorice-scented fennel are a combination to delight both nose and eyes. Some geraniums can be grown in hanging baskets or sprawling over the edges of planters. Stiff, narrow geraniums can be planted close together.
**Species and Varieties:** Rose geranium *(P. graveolens)* has deeply lobed, slightly hairy leaves, and small, pink flowers that bloom in summer. Spreading and more or less erect, it sometimes reaches a height of 4 feet (1.2 m). The species, and two varieties called 'Attar of Roses' and 'Rober's Lemon Rose,' have strong rose scents. 'Elkhorn,' 'Large Leaf Rose,' 'Minor,' and 'Gray Lady Plymouth,' with white-edged leaves, also have a good rose fragrance. 'Red-Flowered Rose' has a fainter scent and bright pink flowers. 'Variegatum' smells like a combination of mint and roses.

Peppermint geranium (*P. tomentosum,* shown in figure 3-36) has velvety, pale green leaves, shaped like grape leaves, and small white flowers. It spreads horizontally and looks good in containers and hanging baskets.

Lemon geranium *(P. crispum)* has several excellent varieties, such as 'Prince Rupert' and 'Prince Rupert Variegated' (alias 'French Lace'). Both have small, wrinkled leaves, lavender flowers, and a stiff, narrow, erect growth habit.

Other fruit-scented geraniums include orange, lime, apple, apricot, and coconut. The best-looking of these are bushy, large-flowered 'Prince of Orange,' apricot-scented 'M. Ninon,' a vigorous plant with bright pink flowers, and lime-scented geranium, *P.* x *nervosum,* with rounded, toothed leaves and good-sized, lavender flower clusters.

There is a good deal of confusion and overlapping among fragrances of spicy-scented and pungent geraniums. "Nutmeg geranium" varieties have small gray leaves, and scents ranging from nutmeg to apple and pepper. Standard nutmeg geranium (*P.* x *fragrans*) is described ambiguously as "spicy-scented"; it has small flowers and long stems, and often grows sparsely. 'Variegatum' is a low-growing variety with irregular white markings on its leaves and a true nutmeg scent. 'Old Spice,' a compact, soft-textured plant, is often apple-scented. 'Rolliston's Unique' is a climbing plant with showy, deep red flowers and a leaf fragrance variously described as pepper, eucalyptus, or mint.

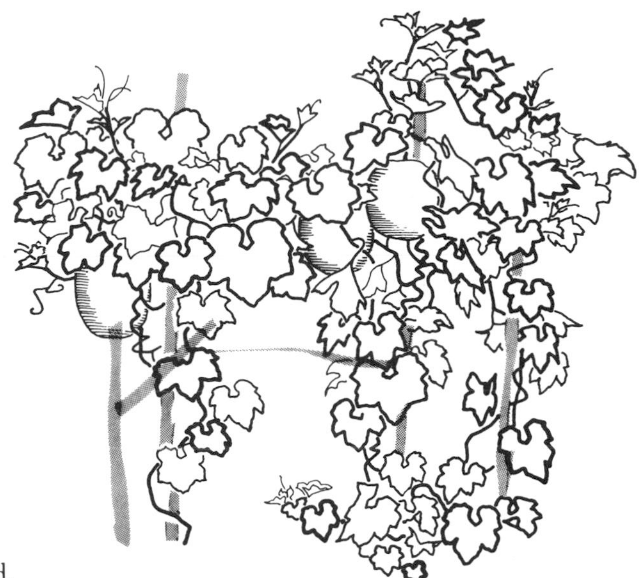

3-37. Fuzzy gourd.

'Fair Ellen' is a good, erect oakleaf geranium, with many lavender flowers and pungent-scented, purple-marked leaves. 'Clorinda' has large, bright pink flowers and a more or less eucalyptus fragrance. Ginger geranium 'Torento' is a big-leaved, bushy plant with lavender flowers and a slight ginger scent. Pine-scented or fernleaf geranium *(P. denticulatum)* has fine-textured leaves and a variable scent—pine or rose. An extremely cut-leaved form, 'Filicifolium,' has a reliable pine scent.

Helen Van Pelt Wilson's *The Joy of Geraniums* (see Bibliography) is a good source for descriptions of scented geranium leaves, flowers, and fragrances. Numerous photographs are in *Exotic Plant Manual,* by Alfred Byrd Graf (see Bibliography). Descriptions of flowering times, plant growth habits, or plant sizes are rare.

Buy scented geraniums in person, so that you can see the plant shapes for yourself and perhaps visit greenhouses where full-sized plants are growing. You can also discover how various geraniums smell to *you*.

## ***Gourd, Fuzzy, and ***Winter Melon (*Benincasa hispida*)
tender annuals

**Ornamental Characteristics:** Fuzzy gourds and winter melons have different fruits, but the plants themselves are very much alike. The climbing vines are attractive and vigorous. Unlike many other plants in the squash family they are single-stemmed, not branching. (See figure 3-37.) They have bright dark green leaves, 6 or 7 inches

(15 or 18 cm) wide, maple-lobed with prominently toothed edges. Their flowers are showy and deep yellow, with parallel veins like those on the bracts of a dogwood. When young, the cylindrical fruits are pale, mottled green, covered with white fuzz. When full grown, winter melon fruits are covered by a thick coat of white wax.
**Ornamental Drawbacks:** The leaves of fuzzy gourd and winter melon may become weather-beaten by late summer.
**Adaptability:** Seed of both plants can be sown early indoors, or outdoors in warm weather. In our trial gardens fuzzy gourd produced fruit in three and one-half months from seed during a cool Oregon summer. It has a reputation for bearing fruit prolifically, but here the crop was small. Warm weather may be needed for a larger crop. Winter melon needs five months of warm weather to produce mature fruit. The fruits should not rest on damp earth or they may rot. When winter melon is grown vertically, its heavy fruits (up to 25 pounds or 11.3 kg) need extra support from slings or platforms.
**Insects and Diseases:** Plants seem very healthy, but watch for squash family pests.
**Food Uses:** Fuzzy gourds, popular vegetables in Asia, are eaten when immature—up to 6 inches (15 cm) long. Chop and add them to curries, bake them stuffed with any of the meats and vegetables familiar in Chinese stir-fries, or use them in the same ways as you would summer squash. Their taste is mild and delicious. Before you cook them, rub them with a paper towel to remove the fuzz, and then peel.

Winter melons can be used like fuzzy gourds when they are young or grown to maturity to be cooked in Chinese soups or stir-fries, or pickled. Mature fruits last a long time in storage.
**Landscape Uses:** Fuzzy gourds and winter melons are attractive additions to most gardens. Their growth is less rampant than that of many of their relatives, so they fit into smaller spaces. They look beautiful growing on a fence, a trellis, or a framework of strings, alone or with flowering vines—morning glory and cardinal creeper. If there is room, they also look pretty rambling along the ground. When planted in a row, they form a central mass about 1½ feet (45 cm) tall and several yards wide, with occasional runners stretching several feet to the sides. These runners can be trained into the central leaf mass or trimmed off to make the plant more compact.

The vibrant green leaves of fuzzy gourds and winter melons show up well against a background of dark evergreen foliage. Marigolds have similar leaf and flower colors and make excellent companions for them. So do yellow and orange cosmos, snapdragon, geranium, dahlia, butterfly weed, and late-blooming lilies and day lilies. You can also train the vines into a mound, bordered formally by neat, uniform plants, such as curly parsley, pepper, marigold, or geranium.
**Varieties:** Fuzzy gourd and winter melon are occasionally available

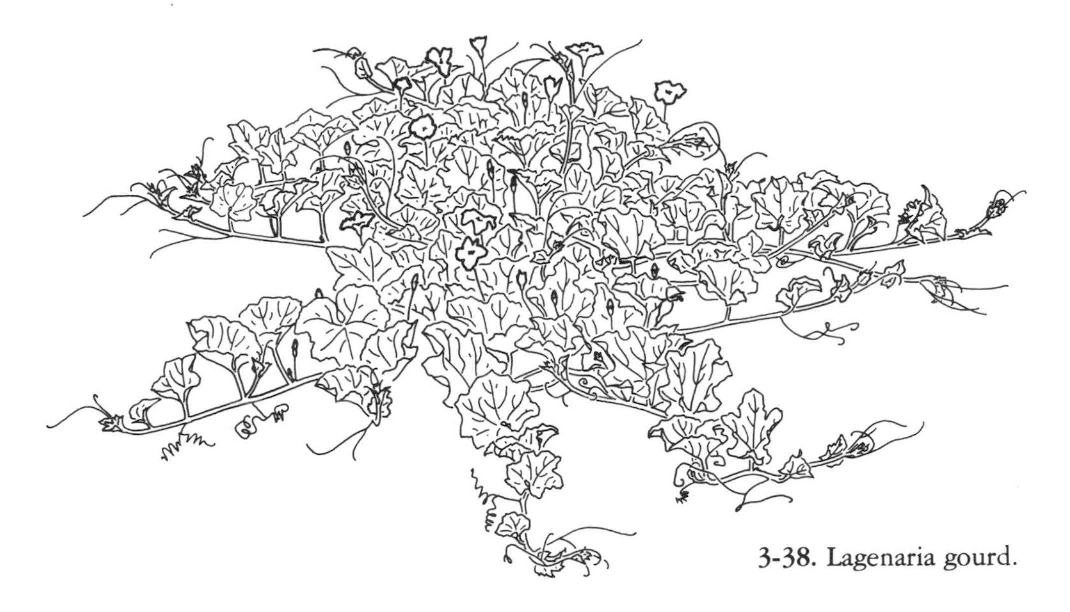

3-38. Lagenaria gourd.

from American catalogs, but no varieties are listed. There are some differences in fuzzy gourd shape according to seed source; some fruits are evenly cylindrical, others taper at one end.

## **Gourd, Lagenaria (*Lagenaria siceraria*)
tender annual

**Ornamental Characteristics:** The lagenaria vine climbs on trees and fences and rambles along the ground, where it is about 2 feet (.6 m) tall and up to 40 feet (12 m) across. Young leaves are lobed; older ones are rounded, with toothed edges that turn up to form a cuplike shape. (See figure 3-38). Leaves are up to 1 foot (30 cm) wide, gray-green, and soft as velvet. Plants are dense at the center, with runners going off in various directions. Floating above the central leaf mass are pale green tendrils and fragrant, white male flowers, up to 4 inches (10 cm) across, on long stalks. Flowers open in the evening and close in the strong light of late morning. Petals are fused and their edges are very wrinkled.
**Ornamental Drawbacks:** Unless lagenaria is grown vertically, its female flowers and fuzzy, pale green fruits of various shapes and sizes are hidden under the foliage. (Fruits turn white and smooth at maturity and yellowish when overmature.) The flowers are not especially attractive when wilted, but there are too many to trim off.

The searching tendrils of lagenaria sometimes give it a wild, tangled look. Lagenaria is very big and not everyone has room to grow it. When bruised, the plant releases a heavy, mildly unpleasant odor.
**Adaptability:** Seed can be started indoors, or planted outdoors in

warm weather. If you start plants indoors, keep them strictly away from any sources of white fly or spider mite.

A long, warm summer increases fruit production. Immature fruits for eating are fairly easy to grow in most areas. It is more difficult to grow hard, mature fruits to use for containers, especially in climates that turn cold or wet early in fall. Growing plants vertically, or putting boards under ripening gourds to lift them off the ground, helps prevent rot in wet weather.

**Insects and Diseases:** Little information is available on this. Watch for squash family pests. Powdery mildew does not seem to be a problem, but plants sometimes get sclerotinia, a disease that causes parts of the vines to brown and die.

**Food Uses:** Immature lagenaria fruits are eaten like summer squash in Italy and parts of Asia and Africa. Some fruits taste mild and pleasant; others are bitter and totally inedible. To be sure of mild flavor, grow only those lagenaria described specifically as food plants. These are listed in seed catalogs as "cucuzzi," "Italian edible gourd," or "New Guinea butter vine." They are usually long-fruited kinds, sometimes grouped under the separate species *L. longissima;* they do not make good dried containers compared to the many other kinds of lagenaria.

For dry containers grow any variety you want, and sample immature fruit to see how it tastes. As fruits grow, they can be wired into various shapes and have designs etched into their surfaces. When mature, they can be dried and waxed, oiled, painted, or carved.

**Landscape Uses:** To prevent rambling lagenaria vines from taking over your garden, you can grow them on trellises, arbors, and trees. They are excellent for hiding chain link fences and tool sheds.

On the ground, the central plant mass is up to 20 feet (6 m) wide and runners ramble out beyond this. You can train runners back into this central mass; the plant will take up half as much space as it would with runners outstretched and it will form a low, broad hedge. The earliest flowers and fruits appear at the plant center. In many climates the runners will not have a chance to bear fruit before frost and you can cut some of them off.

Plant lagenarias where the flowers are visible in the early morning or at night—they look wonderful and ghostly in night lighting. Do not plant them where people frequently brush past, since they have an unpleasant smell when bruised.

The leaves blend nicely with gray-green or blue-green leaves, and with pink, purple, blue, and white flowers. Cosmos 'Radiance' and cleome are good choices. Plants that grow nearby need to be fairly big to hold their own visually.



**Varieties:** Variety names usually refer to the shape or use of mature fruit—bottle gourd, dipper gourd, bird's nest gourd, kettle, pow-

3-39. Grape.

derhorn, and others. Long-fruited types described under "Food Uses" are most often used for eating. Varieties with different fruits are similar in other details of their appearance, although some kinds grow slightly more rampantly than others.

### ***Grape Leaves (and Fruits) (several *Vitis* species) hardy perennials

**Ornamental Characteristics:** Grapes are vigorous, climbing, woody vines with large, maplelike leaves and curling, lacy tendrils. (See figure 3-39.) The green flowers are fragrant, though inconspicuous. The fruit—red, purple, blue, black, green, or yellow—hangs from the vines in lush clusters. The deciduous leaves are an attractive medium green during summer; sometimes they turn red or yellow in autumn. In spring shiny, new, reddish foliage pushes miraculously out of the gnarled, peeling bark of old vines.
**Ornamental Drawbacks:** Some varieties get disfiguring leaf diseases. You need to give grapes plenty of space to grow or resign yourself to regular pruning.

**Adaptability:** You can root your own grape cuttings or buy plants. Vines bear fruit two or three years after planting. They need good air circulation and a steady water supply as fruit ripens. If the crop is heavy or the summer marginally warm, some of the fruit and leafy vine tips should be removed so high-quality fruit will be produced.

To produce heavy crops, grapes should be pruned annually. This is not necessary if you are satisfied with a medium-sized crop and if you grow vigorous varieties well adapted to your climate. A grape grown mostly for ornament may be allowed to spread quickly into a branching system of 400 or 500 square feet (36 to 45 sq m) and trimmed every year to maintain this size. Dead and crowded branches need to be removed from time to time.

Some grapes are disease-prone or fail to ripen fruit in certain climates. Depending on variety, it may take anywhere from 140 to 200 frost-free days to ripen fruit.

**Insects and Diseases:** Many problems can be avoided by planting the right varieties. *V. vinifera* varieties need cool, mild winters and long, hot, dry summers. They grow best west of the Rockies. In humid summers of the eastern states they get downy and powdery mildew. They must be planted on resistant rootstocks, because they are susceptible to nematodes and *Phylloxera* (grape root louse), both prevalent in the eastern United States.

Grapes resistant to Pierce's disease—muscadine grape (*V. rotundifolia* varieties), and some recent introductions of southern experiment stations— are best for the South. In the northern United States, American fox grape (*V. labrusca*) varieties such as 'Concord,' and *V. labrusca* x *V. vinifera* hybrids such as 'Niagara,' do well; they tolerate summer humidity and winter cold.

Other diseases and insects include mildews, black rot, Japanese beetle, and grape berry moth. Birds are often troublesome, too. Many books suggest preventative spray programs for all grapes—three times a year minimum—but in some locations well-adapted varieties are healthy without pesticides.

**Food Uses:** When grape leaves are full-grown but still young and tender, they can be pickled in brine. In Greek and Middle Eastern cooking, meats, vegetables, spices, and rice are rolled inside the leaves, to be served hot or cold. The uses of the fruits are well known.

**Landscape Uses:** Unlimited! Grapes look great climbing a fence in back of a flower-vegetable-herb border. They are an excellent contrast to lacy-textured plants, such as fennel, and to bright flower spikes. Clematis, wisteria, honeysuckle, and kiwi may be strong enough to climb with grape vines, especially if the grapes are pruned more severely than the other plants.



Grapes can cool house walls and patios from hot summer sun. They can shade walkways, screen undesirable views, give privacy, and divide space. Their large, woody branching systems need strong support.

**Species and Varieties:** No information is available on grape varieties with the most tender leaves for brining. *V. vinifera* leaves have been used in the past.

Many new varieties ripen fruit earlier and taste better than familiar 'Concord,' but it is a fine plant anyway—vigorous, attractive, with few insect and disease problems, bearing well even without pruning. Its broad, medium green leaves turn yellow in the fall, and black-purple fruits provide striking contrast. 'Delaware' grapes are described by Robert Nitschke of Southmeadow Fruit Gardens in Michigan as "beautiful ornamental vines with decorative foliage." He considers 'Buffalo' "virtually foolproof—hardy and succeeding with little or no care." Both 'Delaware' and 'Buffalo' have delicious fruits. These varieties are *V. labrusca* stock and *V. labrusca* x *V. vinifera* hybrids. The fruit is used fresh, as juice, and in preserves. *V. rotundifolia* varieties are used in the same ways. All are seedy. Seedless grapes, wine grapes, and some fine, seedy eating grapes all come from *V. vinifera.*

Any grape variety with inherent vigor and good garden conditions looks lush and beautiful, provided it is well adapted to your climate.

Several grape species not ordinarily cultivated for fruit are excellent ornamentals. Japanese *V. coignetiae* has large, rounded leaves that turn red in fall. Native North American grapes include *V. doaniana,* with big, blue-green leaves; *V. vulpina,* with glossy green leaves; and *V. aestivalis. V. vulpina* and *V. aestivalis* have small fruits of good flavor for jelly and juice; so do wild *V. labrusca* plants. For the West, there are two beautiful *V. vinifera* varieties: finely cut 'Apiifolia,' parsley grape, and 'Purpurea,' with its leaves that are red in spring and fall. Little information is available on the edible qualities of some of these species, but it is hard to go wrong. Fruit too tart or seedy for eating raw still makes good jellies and juices.

## *Husk Tomato/Ground Cherry/Strawberry Tomato (*Physalis pruinosa*)
tender annual

**Ornamental Characteristics:** The husk tomato forms a dense, spreading bush up to 2½ feet tall and 8 feet wide (.8 by 2.4m). The heart-shaped leaves are dull green veined with purple, and edged with prominent teeth. (See figure 3-40.) They stay attractive throughout the growing season. Small, bell-shaped, nodding flowers are pale yellow with reddish brown centers. Round fruits, slightly smaller than cherry tomatoes, are enclosed in veined, papery husks. Both fruits and

3-40. Husk tomato.

husks are pale green at first; the fruits turn orange-yellow and husks become tan. As the fruits ripen, they fall off the plants.

**Ornamental Drawbacks:** Plants take up quite a bit of space. Their flowers and fruits cannot be seen from a distance.

**Adaptability:** Husk tomato seeds can be sown early indoors. Germination may be poor and seedlings puny-looking, but when moved outside into open ground, they quickly grow into healthy, vigorous giants. They need full sun and are naturally well adapted to dry, sandy soils. They bear fruit at the same time as late tomatoes.

**Insects and Diseases:** Husk tomatoes do not seem to have the problems that their relatives, tomatoes, often do. Husk tomatoes have few insects or diseases and no physiological peculiarities, such as curling leaves.

**Food Uses:** The skin of husk tomatoes is rather tough, but the pulp is juicy, sweet-and-sour, with the emphasis on sweet—a flavor between that of a cherry tomato and a strawberry or grape, with a strange aftertaste not altogether pleasant. (The aftertaste may disappear when fruits ripen fully.) The husk tomato is eaten raw, alone, or with cream; baked in pies; dried in sugar as a raisin substitute; made into preserves, relishes, and hot sauce; added to salads; fried; baked; and boiled in stews and meat sauce. Husk tomato and other members of its genus are used primarily in Mexican and South American cooking. They are food plants of minor importance in Europe, Asia, and Australia; they were eaten by North American Indian tribes.



Fruits of husk tomato and other edible *Physalis* species should not be eaten until they are ripe. Only the fruits themselves are edible, *not* the husks that surround them.

**Landscape Uses:** You can grow husk tomato far off as a background plant, for its mounds of mild-colored foliage, or plant it in the foreground of a garden where its flowers and fruits can easily be seen. Gardeners who like its flavor and are reluctant to cope with the vagaries of tomatoes in their ornamental plantings may choose to substitute it for tomatoes. However, it is inferior to tomatoes in terms of intense space use; there are no dwarf, bush, or early-bearing varieties available, and I doubt that it responds much to staking or trellising. It could be contained somewhat by growing it in a cage and by using a light hand in fertilizing and watering it.

You might grow husk tomato near other large, bold plants, such as eggplant, unicorn plant, okra, sunflower, zinnia, cleome, and tithonia. Plant it in front of perennials with tall flowering stalks, such as hollyhocks. The purple leaves of perilla and dark amaranthus varieties emphasize the purple veining of husk tomato leaves.

**Species:** Cape gooseberry usually refers to *P. peruviana.* Its other names are poha, ground cherry, winter cherry, and strawberry tomato. It is a tender perennial that resembles *P. pruinosa,* but has larger, bright yellow flowers, and more acid fruits.

Tomatillo or jamberry usually refers to *P. ixocarpa,* a tall perennial with small leaves and large, sticky, purplish fruits that taste best when cooked.

Chinese-lantern plant (*P. alkekengi*) has large, bright orange fruit husks with edible red fruits; it is a common, fast-spreading ornamental that can easily be grown from seed or division, in full sun or partial shade. It can become a weed.

## ***Kale (*Brassica oleracea,* Acephala Group)
very hardy biennial or perennial, often grown as an annual

**Ornamental Characteristics:** Dwarf kale varieties grow in compact, uniform rosettes up to 1 to 1½ feet (30 to 45 cm) tall and 2 feet (60 cm) wide. Leaves are blue-green or gray-green, deeply cut, and tightly curled. (See figure 3-41.) Plants are excellent for winter gardens. During mild winters, they continue to grow; in cold winters, they stop growing in late fall and resume in early spring, producing food until they flower in late spring. Sometimes they are perennial.

**Ornamental Drawbacks:** The yellow flowers are unspectacular. Nondwarf varieties have a loose, tall growth habit; their leaves are less ornate than are dwarf kale's, and they are a decidedly dull green. They have one advantage in winter gardens: leaves are held well off the ground and escape intense ground-level freezes that may harm dwarf kale.



**Adaptability:** Kale seed can be planted in early spring or late summer; sowing in late summer for a fall and winter crop is more common, because kale grows and tastes best in cool weather. However, it

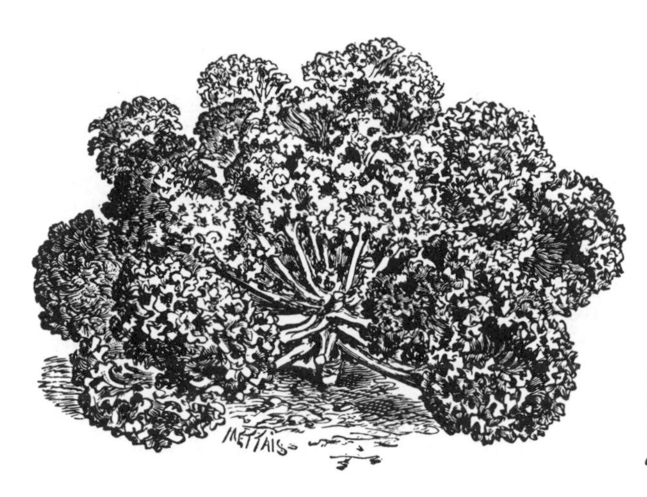

3-41. Kale.
(From Mm. Vilmorin-Andrieux's *The Vegetable Garden.*)

tolerates hot weather as well. It grows best in full sun, but can also live in partial shade.

**Insects and Diseases:** Kale is beset by the same problems as cabbage, but to a much lesser extent. Cabbage worms, aphids, cabbage loopers, harlequin bugs, root maggot, downy mildew, and black leg are all possibilities, especially the first two. Fortunately, cabbage worm holes are relatively inconspicuous in the frilly rosettes of dwarf kale.

**Food Uses:** You can pick off outer leaves of kale and leave the rosette form intact. During winter and spring, more bare stem will gradually appear as the plant grows and you continue harvesting the bottom leaves. Kale leaves have impressive amounts of iron and vitamins A, B, and C. They can be finely chopped and used raw in salads, boiled, steamed, stewed, creamed, baked, stir-fried, or added to soups and potato casseroles.

Eat young kale flowers raw or cooked.

**Landscape Uses:** Dwarf kale plants look great along the edge of a border or path; they make dense foundation plantings. They look fine planted in a mass or in pots and window boxes. You might also grow them near other frilly or fine-textured plants. The dense heads hold their own against big, bold plants as well. They combine nicely with almost any shade of green—bright green, other blue-green or gray-green plants—and with flowers of many colors—white, pink, red, orange, yellow, blue, and purple. In fall they are a mild-colored foil for asters, bright chrysanthemums, and flower kale; in spring they are good companions for spring-flowering bulbs such as tall snowdrop, crocus, star-of-the-snow, daffodil, and hyacinth, as well as for spring-flowering shrubs such as cornelian cherry, winter jasmine, forsythia, star magnolia, and quince. You can plant them around the base of deciduous trees that hold their fruit in winter or bloom early in spring (some crabapples do both) or place them in front of evergreen trees and bushes for a lighter-shaded contrast, alone or interspersed with bright flower kale. Some kale varieties closely match the color of bluish evergreens such as blue spruce.

**Varieties:** 'Dwarf Blue Scotch' ('Dwarf Blue Curled Vates') is an excellent dwarf kale. Deep blue-gray leaves are very curly; plant size and growth habit are quite uniform, and cold-hardiness is excellent. 'Petibor' resembles 'Dwarf Blue Scotch' and has even better uniformity, but, so far, is available only from European catalogs. 'Green Curled Scotch' also resembles 'Dwarf Blue Scotch'; it is almost as nice, but somewhat less cold-hardy. Its height is variable and its gray-green color less interesting than the blue-gray.

'Pentland Brigg,' 'Siberian,' and 'Tall Green Curled' are 2 feet (60 cm) tall or more. They have a more open growth habit than do dwarfs and form long-stemmed, loose rosettes of wavy-edged, gray-green leaves. As rosettes elongate during winter, the plants become awkward-looking and sometimes fall over; 'Tall Green Curled' is the most erect. Plants within a given variety are fairly uniform in size and growth habit. 'Pentland Brigg' has the best cold-hardiness, almost equivalent to that of 'Dwarf Blue Scotch.'

Two European varieties, 'Fribor' and 'Arpad Royal Sluis,' bridge the gap between tall kales and dwarf curly kales. They are tall *and* curly, erect, dense-growing, and very uniform. They are not yet available in U.S. catalogs.

## ***Leek (*Allium ampeloprasum,* Porrum Group) very hardy biennial

**Ornamental Characteristics:** Leek is fountain-shaped and symmetrical, a bold, dramatic form in the garden. Its broad, arched leaves are gray-green, blue-green, or a deep, beautiful blue-gray. They wrap around each other at the base to form a dense white stemlike structure. (See figure 3-42.) Many leek varieties live through very cold winters. In early summer large, fascinating buds appear and then spectacular flowers bloom—spherical lavender balls 6 inches (15 cm) across, each on a stem 3 to 4 feet (.9 to 1.2 m) tall. If some flowerheads are left to ripen and shed seeds, a permanent leek colony will soon establish itself. Leeks also spread by basal offshoots, and, in this sense, they are perennial. The original plant blooms and dies, but the offshoots survive.

**Ornamental Drawbacks:** Leek seedlings grow slowly at first.

**Adaptability:** Leeks do best when seeds are sown in early spring (in a cool, sunny place indoors if the ground cannot yet be worked outdoors). If planted later in spring, they get off to a slower start, but still reach a good size by winter. They thrive in full sun and rich, deep soil with plenty of water, but tolerate partial shade and less than ideal soils.

Leeks take four or five months to reach full size. Some gardeners

3-42. Leek.
(From Mm. Vilmorin-Andrieux's *The Vegetable Garden.*)

hill up the stems with earth (or plant them in trenches and gradually fill in the trenches) to make them white and tender as far up the stems as possible. This blanching process is unnecessary if leeks are cooked for any length of time, because the green portions of stem become tender during cooking. Leeks can be stored outside in the garden without any damage to flavor; they may become tough as flower stalks develop in spring.

**Insects and Diseases:** Leeks are occasionally troubled by some of the same insects and diseases that attack other members of the onion family (see Onion, Bunching later in this chapter) but, for the most part, they are very healthy.

**Food Uses.** You can use delicate-flavored leeks as a substitute for onions or garlic in almost any recipe, but they are subtly different from either one. The tenderest leeks are blanched and used as soon as they mature. You can add them to salads or briefly steam or sauté them. Older leeks must be cooked longer. Quiches and soups are among the best uses for them; they can also be added to stews, boiled, braised, creamed, and stir-fried. They can be frozen or stored in the refrigerator.

**Landscape Uses:** Leeks are best planted where they are fairly inconspicuous during summer, and striking during fall, winter, and spring. They look fine in rows, beds, or winter window boxes. Plant them close together for a dense mass of foliage, or farther apart so the arching leaves have room to spread out.

You might place their bold shapes near bushy plants, such as fennel and sage, or near the distinct rosettes of kale and flower kale. Blue-gray leeks and rose-purple flower kale are an intense combination for late fall and winter. Blue-gray varieties look best near blue-green and gray-green foliage and with many flower colors. They are fine background plants for spring-flowering bulbs—star-of-the-snow, blue hyacinth, daffodil, and iris.

Large-flowered ornamental alliums hold their own and make an attractive combination with leek flowers, as do lilies, day lilies, and the vertical flower spikes of tall lupine, delphinium, and astilbe. Summer-flowering shrubs are good background plants and so are dark green or blue-green coniferous evergreens.

**Varieties:** 'Conqueror' is an excellent variety. It has vigorous growth, deep blue leaf color, and very good winter-hardiness. 'Unique' is also good-looking, but is slightly less cold-hardy. 'Alaska' is an attractive Dutch variety, as hardy as 'Conqueror,' but with less intense blue coloring. 'Broad London' leek, widely available in American catalogs, seems identical to 'American Flag.' Its leaves are less blue than those of other varieties, and its growth is less vigorous. It is somewhat damaged by 12°F. (−11°C.) temperatures.

*A. ampeloprasum* also includes kurrat and elephant garlic. In the eastern Mediterranean, kurrat is cooked as leeks are. It looks similar, but is a smaller plant that does not form the distinctive fountain shape of leeks, and its leaves are narrower. Elephant garlic resembles kurrat; it is grown for its large, mild-flavored, underground cloves.

## *Lemon Balm (*Melissa officinalis*)
hardy perennial

**Ornamental Characteristics:** Lemon balm has softly hairy, oval leaves with scalloped edges and deeply veined surfaces. (See figure 3-43.) In early spring the leaves are glowing green. In full sun it grows in a dense bush, sometimes quite close to the ground—6 inches (15 cm) tall. In shade, plants become up to 18 inches (45 cm) tall and grow more sparsely; they may reach 3 feet (90 cm) in bloom.

**Ornamental Drawbacks:** During summer, lemon balm leaves become duller green. The late summer flowers are white or pinkish and somehow manage to look like old seedheads even in full bloom. Plants die back to the ground in winter.

**Adaptability:** Dry or wet soil, full sun, partial shade, or shade—lemon balm can take it all. Lemon balm spreads by seed and by runners produced after flowering. It is not invasive except in close quarters. To get lemon balm started, you can plant seeds in fall or early spring, make cuttings, or ask neighbors for volunteer seedlings or divisions of clumps in their gardens.

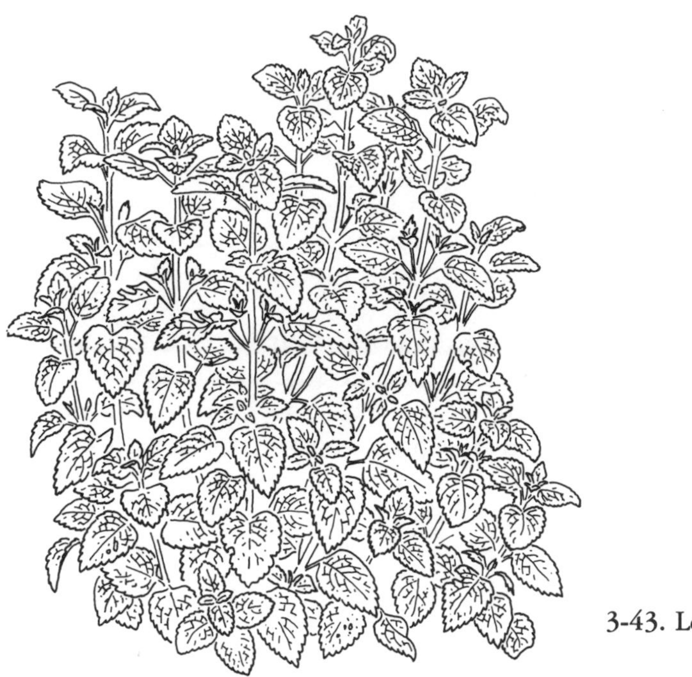

3-43. Lemon balm.

**Insects and Diseases:** Lemon balm is very healthy.
**Food Uses:** The leaves have a mild, slightly lemon flavor and a strong lemon candy scent. They do not taste sweet, but their scent can make them odd additions to nonsweet foods such as tossed salads. They are most often used to flavor fruit juices, fruit salads, canned fruits, punches, wines, and herbal teas. You can add them cautiously to meat, fish, and vegetables.
**Landscape Uses:** Lemon balm plants look attractive growing in small clumps, in masses, or in rows as a bushy spring edging. At their best in early spring, they make an excellent backdrop for spring-flowering bulbs, viola, sweet william, forget-me-not, and spring food plants, such as red and light green lettuce, and thyme and chives (with their spring flowers). They form an attractive ground cover under spring-flowering trees.

Lemon balm, mint, and borage are a vigorous herb combination in partly shaded locations. All of them become more scruffy-looking as summer wears on and should be placed where summer-loving plants will grow up in front of them.
**Varieties:** Leaves of golden lemon balm, *M. officinalis* 'Variegata,' are marked with yellow during spring; in summer they turn uniformly green.



## ***Lettuce (*Lactuca sativa*)
hardy annual

**Ornamental Characteristics:** Colorful lettuce heads range from pale green icebergs and chartreuse butterheads (See figure 3-44) to

3-44. Buttterhead lettuce. (From Mm. Vilmorin-Andrieux's *The Vegetable Garden.*)

bright dark green romaines and deep red-brown loose-leaf varieties. Some have glossy leaf surfaces and others do not. Plant shapes may be tight and ball-like, or loose, rounded masses of leaves. They may be low, neat rosettes or tall, oblong shapes. Leaf margins range from plain to ornate—deeply cut, sharply toothed, and wavy-edged. Leaf surfaces may be smooth, slightly blistered, or very bumpy. Whether the texture of the plants is plain or fancy, they create an effect that is precise and small-scale. Heads seldom exceed 12 inches in height and 18 inches in width (30 by 45 cm).

**Ornamental Drawbacks:** Lettuce is a short-term crop, and most varieties take from six weeks to three months to mature. Part of this time is spent as inconspicuous seedlings, so plants are visually effective for only two to six weeks. Hot weather causes some varieties to bolt or get burned leaves (there are varietal differences in heat tolerance). Some varieties do not develop uniform-sized heads no matter how evenly they are thinned. Some of the best-looking lettuces (loose-leaf varieties in particular) do not taste very good.

**Adaptability:** Lettuce grows best in full sun; it tolerates some shade, but is likely to grow into indistinct leaf masses rather than definite heads. It may become dry and bitter in hot weather or as a result of inadequate water or soil nutrients. In many areas it is sown in early spring or late summer. In moderately cold climates it can be sown late in fall so it will begin growth in spring. In mild climates it is a good winter crop. It tolerates frost down to about 18° F. (−8° C.). 'Arctic King' is a good winter lettuce.

**Insects and Diseases:** Insect pests include cutworms, slugs, aphids, and leaf hoppers. Several diseases are possible, though few cause serious damage; sclerotinia drop, bottom rot, gray mold, mildew, aster yellows virus, brown blight, and tipburn are among the most important. There is some varietal resistance.

**Food Uses:** Lettuce is almost always used fresh and raw in salads, although it is occasionally braised. You can substitute centers of romaine lettuce heads for witloof chicory (Belgian endive).



You can harvest a whole lettuce head at once or just the outer leaves. In times of excess crop, tipburn, or imminent bolting, you may want to eat only the tender central leaves, which can be delicious even when outer leaves are burned or rotting.

Staggered plantings are the most useful way of sowing lettuce. Even though lettuce looks good planted in uniform masses or rows, with seed sown all at once, this method yields heads that will have to be eaten all at once.

**Landscape Uses:** You can plant lettuce in containers, group it in barrels, or line it up in window boxes. If you prefer, it can be massed in the foreground of mixed borders or planted in rows along path edges. Different varieties of lettuce can be combined to create formal geometric designs.

You can sow lettuce fairly thickly and eat the thinnings, maintaining a constant leaf canopy to discourage weeds. Even thinning produces uniform heads in many varieties. Succession planting can be handled in several different ways—plant new lettuce seeds in between older plants at their first or second thinning, start new lettuce in a different location, or sow spring lettuce next to warm-weather vines. When you pull the lettuce out, cucumber, squash, or gourd will ramble over and quickly fill any space that is left.

Red-leaf lettuce looks great with 'Dark Opal' basil, amaranthus 'Molten Fire,' and red-veined 'Rhubarb' chard. You can use the bright green leaves of parsley and green lettuce for contrast against red lettuce. Strong, tough combinations pit red lettuce against the orange of calendula, marigold, and wallflower, or the clear red of tulips and dianthus. Red lettuce and blue-gray leeks or kale are a more subtle duo.

Green lettuce can be grown with other yellow-green or medium green foliage like its own, with red lettuce, or with bright green parsley and celery. It can be interspersed with spring-flowering bulbs, or planted as a ground cover under spring-flowering trees. It looks good in a window box with pansy, viola, heartsease, nemophila, or sweet woodruff.

**Varieties:** Loose-leaf lettuce forms broad rosettes in which the central leaves do not form a ball-like head. It is frequently grown in home gardens, because it grows quickly and you can harvest its leaves over a fairly long period of time. Many heat-resistant varieties with a low tendency to bolting and/or tipburn are included in this group. Some loose-leaf lettuces are thin-leaved and dry-tasting. 'Black-Seeded Simpson' and 'Grand Rapids' are two familiar varieties, with shiny, bright chartreuse leaves that are elaborately wavy-edged and blistered. 'Green Ice' is similar, but darker green, and its leaf edges form definite, rhythmic patterns of wavy lines. It is an especially attractive plant. 'Oak Leaf' and 'Royal Oak Leaf' have narrow, deeply lobed leaves. 'Oak Leaf' is a pleasant apple green, and 'Royal Oak Leaf' is bright medium green at the center of the plant and darker at the edges. 'Salad Bowl' has chartreuse leaves that are lobed, toothed, and wavy-edged in an even, scalloped effect; plants form dense, flattened heads with distinct layers of leaves. 'Oak Leaf' and 'Royal Oak Leaf' have good flavor; 'Green Ice' tastes excellent, even in hot weather. Other varieties taste somewhat dry, though none is very bitter.

Iceberg, or crisphead, is considered the hardest lettuce to grow. It takes a relatively long time until harvest, and does not fare well in hot weather. Fast-maturing varieties have been developed for areas where spring comes late and heats up quickly. Iceberg lettuce forms hard, ball-like central heads surrounded by toothed wrapper leaves. The pale green leaves are thick, juicy, and usually mild-flavored. Main differences in varieties are in uniformity of head size, speed of growth, and ability to hold up well in hot weather. 'Ithaca' is a dependable, uniform variety with excellent resistance to bolting. 'Great Lakes No. 659' and 'Pennlake M.T.' show promise. 'Hot Weather' looks great, but the flavor is not always good.

Romaine, or cos, lettuce forms upright, oval rosettes of dark green, plain-edged leaves with prominent veining. It has a tendency toward dry, strong flavor in its outer leaves. 'Paris Island Cos' and 'Valmaine Cos' are good-looking varieties.

Leaves of butterhead and bibb lettuces are folded into soft central heads. Some varieties such as 'Buttercrunch' M.T.O., dwarf 'Green Mignonette,' and 'Tom Thumb' have crumpled leaves packed tightly on the plants in a way that looks contorted. The leaves of some varieties, such as 'Suzan' and 'Dark Green Boston,' are folded so loosely that the plants look like shapeless leaf piles. 'Summer Bibb,' with greater resistance to tipburn and bolting than regular 'Bibb,' forms uniform, symmetrical rosettes of medium green and is very precise and tidy-looking. 'Butter King' is a fine-looking variety that tastes especially good. Its smooth leaves are shiny apple green and form loose, rounded, rose shapes. 'Butter King' is slow to bolt, though it burns in hot weather.

Red lettuces form different kinds of heads. 'Ruby' is a loose-leaf lettuce of shiny deep red-brown; its leaves are very blistered, wavy-edged, and ornate. Perfect 'Ruby' heads are beautiful, but plants are sometimes very nonuniform. Flavor tends to be dry, strong, and somewhat bitter. 'Salad Trim,' a romaine lettuce with the deep coloring of 'Ruby,' has only slightly better flavor and uniformity. 'Continuity' forms uniform butterhead rosettes, glossy red-brown at the edges and green at the centers, but it tastes dry and bitter. 'Prizehead' has translucent, blistered leaves that are mostly light green, with wavy, red-brown edges. It is a slow-bolting, loose-leaf lettuce that forms uniform, ornate-looking masses of leaves with mild, pleasant flavor.



We grew lettuce in our trial gardens during the summer; varieties described here as dry or bitter may taste much better when grown in very cool weather.

3-45. Lima bean (climbing type).
(From Mm. Vilmorin-Andrieux's *The Vegetable Garden.*)

## **Lima Bean (*Phaseolus lunatus*)
tender, grown as an annual outside the tropics

**Ornamental Characteristics:** Lima beans grow in dense vines (see figure 3-45) or in bushes that are up to 20 inches tall and 40 inches wide (50 by 100 cm). The leaflets, in groups of three, are glossy and wedge-shaped or triangular. Some varieties have very dark gray-green leaves. Others have leaves of contrasting colors—dark and pale gray-green.

**Ornamental Drawbacks:** Bush limas are not very erect; side branches and sometimes whole plants fall over unless supported. The small greenish white flowers and green pods are inconspicuous.

**Adaptability:** Limas are fussy. Soil must be warm for good germination; seed can be started early indoors, or outdoors in late spring. Plants need long, warm summers to produce good crops, but excessive heat results in poor pod set. 'Henderson's Bush' seems the most flexible in its climatic requirements, and 'Fordhook 242' has the greatest ability to set pods at high temperatures. 'Jackson Wonder' and 'Sieva' tolerate heat and drought. In general, baby bush lima varieties bear the earliest crop; pole limas, the latest.

**Insects and Diseases:** These are similar to the ones that attack snap beans. Mexican bean beetles and aphids are frequently encountered. The most likely diseases are downy mildew (on the East Coast of the United States) and anthracnose and nematodes (in the South). Varietal resistance and disease-free seed are good defenses.
**Food Uses:** You can store lima seeds dry for winter use in soups and casseroles. You can also harvest them "green shell," while the pods are still green, but swollen with seeds. (The pods are not eaten.) Young seeds taste sweet; old ones are starchier. Cook green-shell limas fresh, frozen, or canned.
**Landscape Uses:** Plant bush limas in a fairly dense bed so they can hold each other up, and surround them with a small fence, or a hedge of sturdy plants to lean on. This hedge could be a double row of stocky zinnias, geraniums, or dwarf marigolds or perennial hedge of boxwood, southernwood, or germander. Plant climbing limas along a chain link fence, trellis, or string system, with morning glory, moonflower, and cardinal climber. (Give the flowers a head start.) Limas twine loosely and may need to be trained.

Lima leaves go well with green, gray-green, or blue-green leaves, and with many different flower colors. White and blue are especially nice, so you might consider delicate baby's breath and blue salvia. Red-flowered dianthus with gray-green foliage is also attractive.
**Varieties:** There are two main types of limas—one that has big, thick seeds (potato limas), and the other with small, flat, thin seeds (baby limas). 'Henderson's Bush' and 'Jackson Wonder,' both baby lima varieties, have dark gray-green leaves that curl up at the edges. Seed catalogs list them as sixty-five-day crops. 'Fordhook 242' and 'Burpee's Improved Bush Lima,' both listed as seventy-five-day crops, have attractive leaves of two different shades of gray-green. Their seeds are different: those of 'Fordhook 242' are large and thick, while those of Burpee's variety are an intermediate form—large but thin.

The most commonly grown climbing lima is 'King of the Garden,' a dense, loosely twining vine with large, thin seeds.

## **Lovage (*Levisticum officinale*)
hardy perennial

**Ornamental Characteristics:** Lovage is a celerylike plant that grows up to 2 feet (60 cm) tall and 2 to 3 feet (60 to 90 cm) wide its first year from seed. Rather than branching, it forms a dense, vigorous cluster of erect stalks. Dark bluish green leaflets are shaped very much like those of celery, three-lobed and prominently toothed. (See figure 3-46.) In the winter the whole plant dies back to the ground. In very early spring, glossy, reddish shoots push up through the soil, then, in late spring, large yellow flowers bloom on 5-foot (1.5-m) stalks.

3-46. Lovage.

**Ornamental Drawbacks:** Plants look awkward and top-heavy while they are flowering and developing seeds.

**Adaptability:** Lovage is easy to grow from division or seed, in spring or fall. It grows densely in moist soil and full sun, but also does well in partial shade.

**Insects and Diseases:** Lovage leaves sometimes become mottled with brown spots. If stalks with diseased leaves are cut to the ground, healthy new ones soon grow.

**Food Uses:** Leaves, stalks, and whole or ground seeds of lovage are good celerylike seasoning. Use leaves fresh, frozen, or dried; fresh leaves taste delicious in salads. You can blanch the stalk bases and eat them as vegetables. Add roots to soups and stews, either fresh or dried in slices.

**Landscape Uses:** Plant lovage in beds, small clumps, or containers. Its foliage is fresh and strong-colored in spring, when it makes a good companion for small spring bulbs and for tall flowers of late spring and early summer such as angelica, leek, or ornamental allium.

By midsummer, lovage plants often look disheveled and should be in the background of the garden. You can then cut back the stalks after flowering, or leave them while the seeds ripen.

Lovage leaves look very attractive placed near dark green and blue-green foliage—savory, burnet, bunching onion—and blue, white, and purple flowers—nemophilia, forget-me-not, campanula, gypsophila, chives, and crocus.

3-47. Peppermint.

**Species:** Scotch lovage or sea lovage, *L. scoticum,* has similar food uses, shiny leaves, and pink or white flower umbels. Water lovage, *Oenanthe fistulosa,* is considered poisonous.

## ****Mint (various *Mentha* species)**
hardy perennials

**Ornamental Characteristics:** Most mints share certain characteristics—dense growth, fast-spreading roots and pleasantly scented, ovate leaves with even-toothed margins. (See Figure 3-47.) They die back to the ground in winter and send up sturdy, intensely colored new growth the next spring.
**Ornamental Drawbacks:** The summer flowers are often inconspicuous. In addition, some kinds of mint have spreading roots that can become invasive unless contained. Gardeners sometimes grow mint in sunken pots to limit its wandering.
**Adaptability:** Mint does best in rich, moist soil, and full sun or partial shade, but it can adapt itself to a wide range of conditions. Although it can grow in heavy shade, it becomes scrawny and elongated. It can be started in spring from seed or cuttings, but division is the easiest, fastest way. Neighbors often have plenty of extra mint to share. Mint plantings need to be renewed after about four years, because, as plants spread outward, old stems in the center die.
**Insects and Diseases:** Rust and slugs are sometimes problems.

**Food Uses:** Leaves of various mints, especially spearmint, are used to garnish and flavor teas, juices, cocktails, salads, fruits, desserts, meats, vegetables, rice, jelly, vinegar, chutney, and sauces. Peppermint and to some extent spearmint are used in herbal teas to relieve indigestion and fever.

**Landscape Uses:** Mint is an excellent ground cover in the dappled shade of light-canopied trees and shrubs. Various kinds of mint can be used alone or combined with other sturdy ground covers, preferably those that stay green in winter when mint dies back. Bugleweed is a good choice.

Mint can grow in containers. A pot of mint sunken in the ground under a water faucet will thrive and look pretty in moist conditions. Corsican mint forms a low, mosslike mat that can grow over rocks, along a garden pool, or between the stones of a path.

Mint leaves, usually dark bright green, go well with other bright greens—lemon balm, parsley, celery—and with blue-green fennel, bunching onion, and dwarf juniper. They are especially vivid early in the growing season, when they make an excellent background for bright flowers of spring and early summer. Late in summer mints look drab and belong in the background.

**Species and Varieties:** Spearmint, peppermint, orange mint, apple mint, pineapple mint, red mint, golden mint, and Corsican mint are all good ornamentals. Most are excellent for general culinary use, except for Corsican mint (Crème de menthe mint), which is traditionally used as a liqueur flavoring.

Spearmint (*M. spicata*) has erect stems from 1 to 3 feet (30 to 90 cm) tall, bright green, ovate leaves with sharply toothed edges, and slender spikes of pink or lavender flowers. Its creeping roots spread quickly.

Peppermint (*M.* x *piperita*), spreads by both roots and runners. Its 1- to 3-foot (30- to 90-cm) stems may be erect or spreading; flower spikes are lavender. Leaves are darker green than those of spearmint. "Black" peppermint has purplish coloring in leaves and stems; "white" peppermint is completely green. The oval flower spikes are lavender.

Orange mint (*M.* x *piperita* var. *citrata*) has dark green, heart-shaped leaves with purple spots and/or leaf margins, and a pleasant citrus fragrance. Its reddish stems are up to 1 foot (30 cm) tall, and not at all erect. Flowers bloom in light pink, oval spikes.

Apple mint (*M. suaveolens*) has erect stems 2½ feet (75 cm) tall, and sweet-scented, rounded, wooly leaves, gray-green on top and white underneath. Dense, cone-shaped flower spikes are pinkish white. Pineapple mint, the apple mint variety 'Variegata,' is only 1 foot (30 cm) tall. Its small leaves are blotched with white. Young foliage in particular is clearly pineapple-scented.



Red mint (*M.* x *gentilis*) has toothed, dark green leaves that are

3-48. Mizuna.

spearmint-scented and variable in shape and hairiness. It has 1- to 1½-foot (30- to 45-cm) stems, erect and reddish purple. Pink flower clusters bloom up and down the stems in leaf axils. Golden mint (*M.* x *gentilis* var. *variegata*) has leaves splotched with yellow or white and spreading stems.

Corsican mint (*M. requienii*) is a 1-inch (2.5-cm) ground cover with tiny, round green leaves and sparse pink flowers that are reliably hardy to only 5° F. (−15° C.).

## **Mizuna (*Brassica japonica*)
hardy biennial

**Ornamental Characteristics:** Mizuna grows in uniform, vigorous rosettes up to 16 inches tall and 2 feet wide (40 by 60 cm). Leaves are bright, glossy green, veined with white, finely divided, and edged with prominent teeth that give the plants a jagged, ornate appearance.
**Ornamental Drawbacks:** I am not aware of any. Mizuna is an attractive, reliable plant.
**Adaptability:** Mizuna germinates well even in cold soil and can be planted from early spring to early fall. It tolerates a good deal of hot weather. In cool summers it may live throughout the growing season from an early spring planting. It can take some frost. Even when leaves are killed back in winter, the crowns may survive to send up new growth in spring.
**Insects and Diseases:** Mizuna seems to have little susceptibility to pests that plague its cabbage family relatives.
**Food Uses:** Mizuna tastes best during its first two months of

3-49. Mustard greens.

growth. You can make succession plantings for food use and leave older plants in the garden for purely ornamental purposes.

You can add mizuna to soups and stir-fries. In Japan it is very popular; its mild taste blends easily with other foods.

**Landscape Uses:** Grow mizuna in containers, in beds, or in rows, as a neat edging for borders and paths. It is an excellent background plant for low-growing flowers of spring, summer, and fall; it goes well with green and blue-green foliage.

**Varieties:** A plainer-leaved, less attractive variety of mizuna is available in Japan, but as far as I know it is not available in the United States.

## **Mustard Greens (*Brassica juncea*)
hardy annual

**Ornamental Characteristics:** Mustard leaves are very bright green, reaching flourescent intensity at the leaf edges. Edges may be toothed and rather plain, or cut, fringed, and wavy. Plants grow in dense, upright masses that fan outward at the edges and have good uniformity. They grow up to 2 feet high and 2 feet wide (60 by 60 cm).

**Ornamental Drawbacks:** Once flowering begins, the plants are no longer attractive.



**Adaptability:** Fall and very early spring are the best times to plant mustard greens. In fall the plants last longer. They are sensitive to heat and dry soil and respond quickly by bolting. Mustard greens can survive moderate frosts.

**Insects and Diseases:** Mustard is less susceptible to insect and disease attack than many of its cabbage family relatives. Possible problems include cabbage worms, aphids, harlequin bugs, root maggots, downy mildew, and black leg.

**Food Uses:** Use vitamin-rich mustard greens raw in salads, cook them alone or with other greens, or add them to soups, stews, and stir-fries. They taste best when young, and become more pungent and tough as they age. Cooking, however, softens them and makes them taste milder. You can harvest lower leaves throughout mild winters.

You can use the red-brown seeds whole to season meats or grind them to make hot mustard.

**Landscape Uses:** Grow mustard greens in containers, beds, or rows. They are boldly effective because of their color and size and must be used thoughtfully. Quite a few colors clash with glaring green, but it can look outstanding in the right place.

For example, mustard contrasts well with very dark broad-leaf and coniferous evergreens. It stands out clearly against a plain background of wood or gray stone and it can be planted with early spring bulbs or fall chrysanthemums, especially yellow and white ones, or with bright green parsley, celery, mizuna, and lettuce varieties.

**Species and Varieties:** 'Prizewinner' is a very attractive plant. Its erect leaves are fairly plain with toothed, curly, bright green edges; they are a dark intense green at their centers. 'Green Wave' is similar and almost equally attractive. Its leaves are somewhat less erect and its coloring less vivid. Both varieties taste very hot once they grow big.

Two other good-looking varieties, 'Fordhook Fancy' and 'Southern Giant Curled,' have leaves that are cut along the edges and extensively curled. They are quite erect in growth habit. 'Fordhook Fancy' becomes as pungent as 'Prizewinner' and 'Green Wave;' 'Southern Curled Giant' is moderately hot.

'Florida Broadleaf' and 'Shirona' are plainer-leaved and less erect than other varieties; 'Shirona' has a relatively mild flavor. In our trial gardens we grew 'Miike Giant' mustard, probably a form of gai choy or Chinese mustard; it has crumpled green leaves with muddy, reddish purple coloration and looks like a squashed Swiss chard.

'Tendergreen,' spinach mustard or mustard spinach (*B. rapa,* Perviridis Group), has good resistance to bolting. It is a plain-leaved plant, less showy than *B. juncea,* but mild and pleasant-tasting even when it grows big.

Black mustard (*B. nigra*) and white mustard (*B. hirta*) are grown for mustard seed. Young leaves of both species are sometimes used as greens. Black mustard and wild mustard (*B. kaber*) grow wild through-

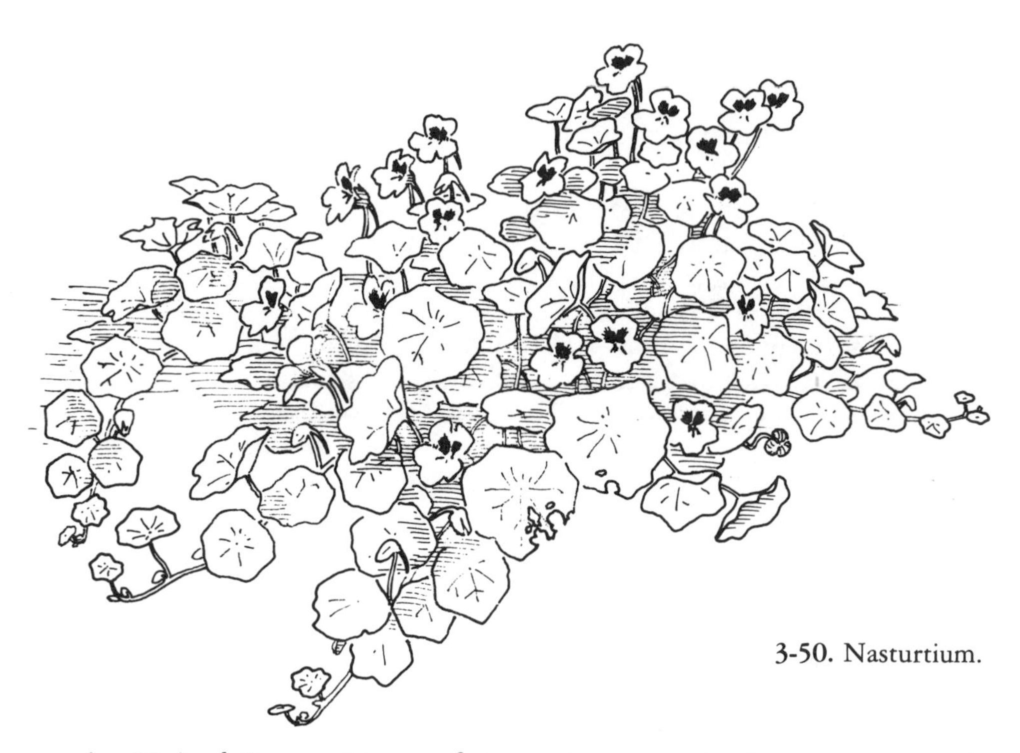

3-50. Nasturtium.

out the United States. None of these three species have ornamental possibilities.

### **Nasturtium (*Tropaeolum majus* and *T. minus*)
tender annual, perennial in frost-free areas

**Ornamental Characteristics:** Nasturtium leaves are smooth and round with slightly scalloped edges. (See figure 3-50.) They are pale gray-green or blue-green, with whitish veins that come from the center of each leaf like spokes on a wheel. Growth habits are variable. There are bush and dwarf bush varieties, derived mostly from *T. minus,* and also climbing and trailing forms derived from *T. majus.*

Nasturtiums bloom within two months from seed. The narrow buds and wide-open flowers are vividly colored and gracefully shaped. Spurs are filled with nectar, and hummingbirds may come to sip it. Flowers are often fragrant and either single or double. The colors range from pale yellow through flaming orange to deep mahogany red. Sometimes the petals are a contrasting color at the throat. The plump fruits are also attractive. Plants sometimes reseed themselves and form a long-lasting nasturtium patch.

**Ornamental Drawbacks:** Nasturtiums vary a great deal with garden conditions. In rich soil, they produce large, long-stemmed leaves that hide the flowers. Double flowers often look contorted. The most frequent color combination among nasturtiums is bright orange-red flowers with pale gray-green leaves. These colors are hard to combine visually with other plants.



Growth habits are hard to predict from descriptions in many seed catalogs. Some "climbing" varieties would much rather be trailing plants and have to be coerced into climbing. Many "bush" varieties are actually spreading ground covers.

The biggest ornamental drawback is black aphids. You may turn over a nasturtium's pale leaves and find that they are crawling with wall-to-wall aphids—a solid, slow-moving mass. New leaves may become stunted, but aphid damage is seldom extensive. It is their presence alone that is decidedly nonornamental.

**Adaptability:** Nasturtiums produce the greatest numbers of flowers during cool summers, in dry, gravelly, or sandy soil, and full sun. With shade, moisture, and/or rich soil, plants grow profuse leaves, fewer flowers, and are effective mainly as foliage.

It is best to sow nasturtium seeds in place in late spring. Seedlings can be transplanted, but take a while to recover afterward.

**Insects and Diseases:** Possible problems are aphids, flea beetles, cabbage worms, slugs, snails, leaf miners, and wilt.

**Food Uses:** Leaves, stems, flowers, and young fruits taste juicy and peppery. Eat all of them fresh in salads and sandwiches, or pickle them. The buds and fruits are a good double for capers. Stuff the flowers or combine them with other edible flowers to make flower vinegar or flower pickles.

**Landscape Uses:** You can grow nasturtiums in raised planters or on terraces. At eye level, their flowers are visible no matter what the growth habit of the variety. Grow dwarf bush nasturtiums in a mass, or as edging for a path or garden. They make an attractive border for a water lily pool. Large "bush" nasturtiums, and climbing or trailing nasturtiums left to ramble along the ground, are excellent ground covers. You can use trailing nasturtiums in window boxes, hanging baskets, and other containers; large trailers may hang several feet down a wall. Climbing nasturtiums twine loosely. They climb best on wire mesh or string mesh, and may need help.

You can combine the pastel gray-green of most nasturtium leaves with bright green leaves of plants such as parsley and celery, or with blue-green. Rue has a color close to that of nasturtiums. Finely cut leaves are a good contrast to their plain outlines. Nasturtiums are often grown with other flowering plants, but seldom to visual advantage; it is easier to combine them with foliage plants.

**Species and Varieties:** In the rich, moist soil of our trial gardens, any tendencies toward large, long-stemmed leaves and rampant growth showed up clearly. Most varieties grew about 2 feet (60 cm) tall, with varying widths.

'Jewel Mixture' grew 5 feet (1.5 m) wide, with dense, long-stemmed foliage that hid its nicely colored double flowers.

'Single Tall Climbing Mixed' became 10 feet (3 m) wide and bore a few single flowers with a narrow color range. The flowers of 'Dwarf

3-51. Okra.

Dark-leaved Mixed', 6½ feet (2 m) wide, had a wide range of colors, but were mostly hidden under dense, attractive mounds of foliage—blue-green on some plants, pale gray-green on others.

'Gleam Mixture' and 'Empress of India' are especially attractive nasturtiums. 'Empress of India' has bright red single flowers and blue-green foliage that is purplish when young. 'Gleam Mixture' may grow almost 8 feet (2.4 m) wide. Although leaf stems are long and flower stems are short, growth is fairly open, so the many double flowers are clearly visible. Their color range seems narrow—mostly school bus yellow and dark orange-red. Leaves are especially large, some over 5 inches (13 cm) in diameter.

South American mountain people eat the lumpy tubers of ysano or tuberous nasturtium (*T. tuberosum*). Plants are vigorous, perennial vines that may grow 3 to 10 feet (.9 to 3 m) long. Leaves are lobed, and small, plentiful orange and yellow flowers are shaped like narrow tubes. Overwintering tubers are hardy to 20° F. (−7°C.).

## ***Okra/Gumbo (*Abelmoschus esculentus*)
tender annual, perennial in frost-free areas

**Ornamental Characteristics:** Okra has toothed, maple-lobed leaves of medium green and pale yellow funnel-shaped flowers with deep red throats. (See figure 3-51.) The red, green, or white fruits are narrow, ridged cylinders with pointed tips. In good conditions, okra grows into a vigorous, well-branched shrub, up to 8 feet (2.4 m) tall, depending on variety.



**Ornamental Drawbacks:** In cool summers, okra plants are small and scrawny. Leaf diseases that sometimes affect them are disfiguring. The beautiful flowers last for only one day.

**Adaptability:** Okra seed germinates slowly and benefits from a day of soaking before it is planted in warm soil. When transplanting seedlings, keep root disturbance to a minimum. Okra needs warm summers with warm nights; in cool summers it does poorly, even when grown on black plastic.

**Insects and Diseases:** Possibilities include bollworms, stink bugs, leaf beetles, corn earworms, nematodes, and fusarium wilt.

**Food Uses:** Okra pods grow fast; you must harvest them every couple of days. You can boil, fry, deep-fry, bake, or add them to soups and stews. (Their mucilaginous quality sometimes verges over into sliminess, which you can reduce by handling the pods carefully to prevent bruises and breakage and by cooking them quickly.) You might also can, freeze, dry, or pickle them in brine.

**Landscape Uses:** Okra is an effective background plant used alone or with other big plants such as corn, hollyhock, or roses. Plant large-scale ground covers, such as squash, gourd, or cucumber, in front of it. You can use okra as a tall hedge or foundation planting; staking may be helpful in exposed locations.

**Varieties:** Red okra is the most striking. Its green leaves are veined and splotched with deep crimson; stems, flower buds, and fruits are also deep crimson. Compared to the flowers of green okra varieties, its yellow flowers have especially strong red coloration in their throats and on their stigmas.

Green varieties include 'Emerald,' 'Clemson Spineless,' and 'Dwarf Long Pod Green.' None does very well in cool weather, but 'Emerald' is more vigorous than others. All bear fruit about two months from seed.

## ***Onion, Bunching/Welsh Onion (*Allium fistulosum, A. fistulosum* x *A. cepa* hybrids, and some *A. cepa* varieties)

very hardy perennial clumps

**Ornamental Characteristics:** Bunching onions are quite different from common onions. Common onions (*A. cepa*) listed under "Less Attractive Vegetables and Herbs" at the end of this chapter, have underground bulbs and thin, spindly leaves that often fall over. Bunching onions seldom bulb; they are splendid ornamentals. They have tall, hollow leaves, round in cross section and pointed at the tips. They grow in spiky, vertical masses, 3 feet tall and 1½ feet wide (90 by 45 cm) the first year from seed. (See figure 3-52.) Their effect is stark and bold; leaves endure a good deal of frost and often last through winter. Most are blue-green; some are medium green. New shoots develop

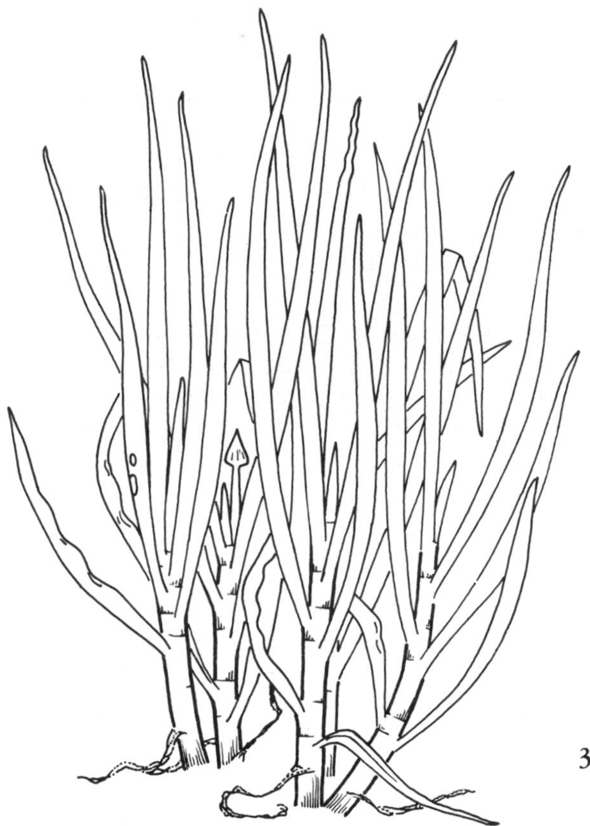

3-52. Bunching onion.

and leaf masses grow wider through the seasons. In spring they bear round yellowish or white flowerheads, but individual florets on the head do not always open.

**Ornamental Drawbacks:** Although interesting, the flowerheads are not very pretty, and, in preparation for a brief resting period during early summer, foliage dies down as flowers bloom. A few varieties have floppy or twisted leaves and little vigor. Some varieties split into awkward-looking clumps of leaves leaning out in various directions.

**Adaptability:** Germination is sometimes erratic, so sow seed thickly in loose-textured soil during early spring or late fall. Although bunching onions grow most vigorously in full sun and rich, moist, well-drained soil, they adapt to less ideal conditions as well. You should divide and move them into fresh soil every three years. They often reseed themselves.

**Insects and Diseases:** Bunching onions have few insect or disease problems. Onion thrips and maggots sometimes bother them, but they are resistant to most diseases of common onions.

**Food Uses:** In most climates, except for a brief rest after flowering and before new leaf growth begins, bunching onions can be used as scallions on a year-round basis. Leaves become stronger-tasting as they grow and are tough-textured just before flowering.



**Landscape Uses:** The leaves are effective in masses or clumps, and

3-53. Rocambole (serpent garlic) just before flowering.

in rows as edging or foundation plants. They look great with other blue-green foliage of contrasting texture, such as fennel, kale, or savoy cabbage. They also combine well with bright green or gray-green foliage, and with blue, purple, or pink flowers. Flower kale and bunching onions make excellent mild-winter companions. You can place bunching onions in back of early spring-flowering bulbs or small fall asters.
**Species and Varieties:** 'Ishikura Long Bunching' and 'Kujo Green' are very erect, vigorous varieties. 'Kujo Green' has especially intense blue-green coloring. 'Japanese Bunching,' 'Common Bunching,' and 'Evergreen Long White Bunching' are attractive, dependable varieties with only slightly less vigor and erectness than the first two.

'Beltsville Bunching,' 'Perfecto Blanco,' and 'White Bunching' (='White Lisbon') develop small bulbs, which force their leaves to lean out at odd angles, rather than forming a single, erect mass.

Several other *Allium* species have good possibilities as ornamental food plants. Nodding onion (*A. cernuum*) blooms in summer, with rose or white flowers in loose, nodding heads. Its 2-foot (60-cm) leaves are narrow and erect. Native to most of North America, it grows and spreads easily in dry soil. Its strong-flavored bulbs can be used in soups and pickles.

*A. senescens* bears pink flower umbels in early summer. Leaves may be straight or twisted, green or gray-green, and plants are about 6 inches (15 cm) tall. The botanical variety *glaucum* has gracefully twisting blue-gray leaves and late summer flowers. In Japan the bulbs are sometimes eaten.

*A. sphaerocephalum* has grasslike leaves and spherical umbels of dark purple flowers that bloom in midsummer. It grows 2 feet (60 cm) tall and the bulbs are used as food in Siberia.

Several other *Allium* species, less distinguished as ornamentals, are still valuable food plants. Garlic (*A. sativum*) has thin, sparse leaves. Shallot (*A. cepa* Aggregatum Group) has dense clumps of narrow, dark green leaves. Bulbs are planted in winter or early spring. They multiply quickly and you can harvest them when leaves die back in July. Bulbs and leaves are delicately onion-flavored.

*A. cepa* Proliferum Group, known variously as Egyptian onion, tree onion, or top onion, produces underground bulbs and also clusters of small bulbs at the top of its stems. Rocambole or serpent garlic (a form of *A. sativum*) has flat, leeklike leaves and flower stems that twist into serpentine coils before they straighten to produce clusters of small, garlic-flavored bulbils at their tops. (See figure 3-53.) Leaves, bulbs, and bulbils of Egyptian top onion and rocambole are edible, but neither plant is especially attractive.

## **Oregano/Wild Marjoram (*Origanum vulgare*)
hardy perennial

**Ornamental Characteristics:** In full sun, oregano is a dense, low ground cover 2 inches (5 cm) tall. In partial shade it is an upright bush about 1 foot (30 cm) tall. Its soft, hairy leaves are heart-shaped, strongly veined, and medium green with a yellowish cast. Stems are purplish and woody. (See figure 3-54.) The long-lasting, late-summer flower clusters are dark purple when in bud and lavender-pink when open. They are carried on waving, semi-erect stems.

Oregano stems root as they grow; plants grow up to 1½ feet (45 cm) across the first year from seed. Many of the leaves turn brown in winter, and, in cold climates, plants may die back to the ground. They quickly become lush and vigorous again, however, in spring.

**Ornamental Drawbacks:** The flowering stems of oregano sometimes look messy; they become elongated when plants are growing in partial shade.

**Adaptability:** Oregano is a vigorous plant adapted to full sun or partial shade. It does well in poor, dry soil and grows even better in fertile soil. It has good cold-hardiness. Seeds can be sown in early spring or you can start plants from cuttings or division. (They can be transplanted easily.)

**Insects and Diseases:** Oregano is free of insects and diseases.

**Food Uses:** Oregano is mild-flavored and you can use it liberally as a seasoning, either dried or fresh. (Dried oregano available commercially is often a stronger-flavored Greek form; see "Species and Varieties.") Many people consider the flavor of oregano inferior to that of sweet marjoram.

3-54. Oregano.

Oregano tea is sometimes used as a headache remedy. In India, the leaves are cooked as potherbs.

**Landscape Uses:** Oregano is an excellent ground cover that is especially useful on banks and eroded slopes where its strong roots hold the soil well. Use it as a ground cover by itself or with creeping thyme; you might also use it as lawn edging, since it can be mowed and holds its own against grass. It can be planted in clumps or large masses, or in rows as an edging plant. It can be combined with small spring flowers, or white, pink, blue, or purple flowers of late summer, such as heliotrope and verbena. Its texture and colors blend with a wide range of other plants.

**Species and Varieties:** Only the species is available from seed catalogs, but herb nurseries sell the variety 'Aureum,' which grows in neat mounds of yellow-green leaves.

Greek oregano (*O. vulgare 'Viride'*) is less cold-hardy than the species; its leaves are strong-flavored and its flowers are white. You can buy it from herb nurseries or grow it from seed; a good seed source is packaged flowering tops of oregano imported from Greece (sometimes sold in imported-food stores).

Marjoram or sweet marjoram (*O. majorana*) appears under "Less Attractive Vegetables and Herbs" at the end of this chapter. Several other *Origanum* species are grown infrequently in the United States. Pot marjoram (*O. onites*) resembles sweet marjoram but is more cold-hardy. Pot marjoram or winter sweet marjoram (*O. heracleoticum* ) resembles oregano, but has a stronger flavor.

3-55. Curly parsley. (From Mm. Vilmorin-Andrieux's *The Vegetable Garden.*)

Dittany of Crete or hop marjoram (*O. dictamnus*) has rounded gray leaves and prominent pink flower clusters. It tolerates little cold and is usually grown as a potted plant that spends winter indoors. It is used as seasoning for food or for vermouth.

## ***Parsley (*Petroselinum crispum*)
hardy biennial often grown as an annual

**Ornamental Characteristics:** Parsley forms neat, rounded clumps from 6 inches to 1½ feet (15 to 45 cm) tall and from 1 to 2½ feet (30 to 75 cm) wide. Leaves are a vivid emerald green, most intense on new growth. They are finely divided into toothed segments. Leaf surfaces may be thoroughly crinkled (curly parsley) or flat (Italian and Hamburg parsley). Curly parsleys (See figure 3-55) have a crisp, stiff look because of their foliage and are smaller than flat-leaved plants, which have a loose, fernlike appearance and softer green coloring.

Parsley tolerates some frost (see figure 3-56) and provides fresh color in the fall garden. During winter, leaves may be killed by hard frosts, but roots often endure to send up bright shoots in early spring.
**Ornamental Drawbacks:** In late spring of the second year, parsley plants flower, with unattractive greenish umbels on 2- to 3-foot (60- to 90-cm) stems. They can reseed themselves to form an ongoing parsley patch, but you need an inconspicuous place for them to do it.

Some varieties are uniform in size and others are not. This matters only if you are using the plants for edging.
**Adaptability:** You can plant seeds in early spring. Germination takes several weeks and requires loose, moist soil. If your soil tends to crust at the surface, cover it with newspaper or plastic until the parsley seeds begin to germinate, or add some looser medium—sand, peat moss, etc.—to cover them. You can start seeds indoors if you transplant carefully to avoid damage to the seedlings' taproots.

3-56. 'Dark Green Italian' parsley on a frosty morning.

Parsley grows slowly at first and comes to full glory in late summer, responding enthusiastically to moist, fertile soil. It thrives in full sun as long as soil does not dry out, and does well in partial shade.

**Insects and Diseases:** Parsley is a vibrantly healthy plant.

**Food Uses:** You can clip parsley's outer leaves frequently without hurting the plant's appearance, as long as you cut leafstalks at ground level without any stubs left sticking up.

Parsley is rich in iron and vitamins A and C. Use it fresh, dried, or frozen, in salads; in salad dressings, sauces, and garnishes; and as seasonings for vegetables, meat, and rice (usually in small quantities, although some Middle Eastern and Indian recipes use large quantities). You can substitute parsley for basil in making pesto.

Both leaves and roots of Hamburg (turnip-rooted) parsley provide food, but harvest the tops sparingly. Leave the roots in the garden until you need them and use them as you would carrots.

**Landscape Uses:** Parsley looks wonderful in borders, edgings, masses, planters, pots, and window boxes. It makes a fine combination with white, yellow, orange, red, blue, and purple flowers; pink and lavender are not as good. Bright green, dark green, yellow-green, and blue-green are all attractive foliage blends; gray-green doesn't work well. Fine-textured foliage plants, such as carrot and coriander, match parsley, and small plants with large leaves (romaine lettuce, for example) are good contrasts. Large, bold-textured plants are overwhelming.



Some effective combinations include curly parsley with lobelia,

3-57. Parsnip.

'Lulu' marigold, viola, or linaria; flat-leaved parsley with kale and gysophila; any parsley with small spring-flowering bulbs, such as dwarf narcissus and grape hyacinth. As an edging plant, parsley can be used alone or combined with other bushy, long-lasting plants such as dwarf pepper and dwarf marigold.

**Varieties:** Many varieties are good ornamentals. Some are very uniform in size and a few have outstanding vigor. 'Dark Green Italian,' 'Banquet,' and 'Deep Green' (the latter two are curly-leaved) are great-looking varieties. 'Darki,' 'Extra Curled Dwarf,' 'Paramount' (all curly-leaved), 'Plain Hardy Italian,' and 'Hamburg' are also very good. 'Extra Triple Curled' has a conspicuous lack of uniformity.

## **Parsnip (*Pastinaca sativa*)
hardy biennial grown as an annual

**Ornamental Characteristics:** Parsnip plants form attractive fountain shapes, 1 to 1½ feet (30 to 45 cm) tall and 1½ to 2½ feet (45 to 75 cm) wide. The large, bright green leaflets are lobed and toothed. Parsnip resembles celery, but its leaf texture is bolder. (See figure 3-57.)

Leaves last far into fall —and in mild climates they last all through winter. When leaves are killed by cold weather, new ones sprout in spring from any roots left in the garden. Big yellow-green umbels bloom in late spring at the top of 3- to 6-foot (.9- to 1.8-m) stalks.

**Ornamental Drawbacks:** Parsnip's leaf stalks sometimes bend so much that plants become loose and floppy-looking. However, when closely planted, they hold each other up.
**Adaptability:** Parsnip seeds have short viability, and fresh seed should be used each year. Germination is slow and uneven. Plant early and thickly, in loose, moist soil. If your garden soil crusts at the surface, proceed as for parsley.

Young seedlings are slow-growing and unable to compete with weeds, but older plants need little attention. They have a long season to harvest—up to four months—and you can leave them in the garden until you are ready to use them. Mulching in the fall protects roots from cracking when the soil alternately freezes and thaws during winter.
**Insects and Diseases:** Insects and diseases that affect carrots may also affect parsnips.
**Food Uses:** Temperatures near or below freezing make parsnips taste sweeter, so leave them in the ground to improve their flavor. Steam fresh or frozen parsnips, boil them, cook and then purée them, or add them to stews and soups.
**Landscape Uses:** Combine parsnip foliage with a broad spectrum of colors—white, yellow, orange, red, blue, purple, blue-green, bright green, dark green, and yellow-green—and with other intermediate leaf textures—celery, loosely curled kales—or contrast it with fine or bold textures. You can grow parsnip plants in rows or masses, in raised beds or in the open ground; most varieties are too deep-rooted for pots or window boxes.

Parsnips are good foreground plants for fall gardens, along with chrysanthemums, cabbage, and white flower kale. They provide bright color during mild winters, and in spring they make a good background for spring-flowering bulbs. In late spring the flower umbels are striking in a tall mass behind other plants.
**Varieties:** 'Harris Early Model' grows larger than 'Improved Hollow Crown' and has sweeter-tasting roots. 'Early Round' is a short-rooted variety suitable for shallow or heavy soils.

## *Pea (*Pisum sativum*)
hardy annual

**Ornamental Characteristics:** The curling tendrils and oval leaves of peas are attractive in a quiet, unspectacular way. Leaves are soft dark green with whitish mottling. Flowers look like small sweet pea flowers. They may be either white or lavender-pink. As tall pea varieties climb a trellis, they form a dense, vigorous wall of pleasant flowers, foliage, and tendrils.

3-58. Pea.

**Ornamental Drawbacks:** "Bush" pea varieties have a messy growth habit. Small vines rather than true bushes, they lean and wrap their tendrils around each other and soon form a tangled mass of interwoven stems. Flowers are hard to see in the confusion. Some varieties turn out to be taller than their catalog descriptions and you may have to stake them.

**Adaptability:** Peas grow best in cool weather; early spring plantings are the most successful in most locations. In very mild climates, you can grow peas during winter—they tolerate several degrees of frost. Where winters are slightly colder, they can be sown in late fall for early growth the following spring.



Garden soil for peas should not have excess nitrogen, which causes rank leaf growth at the expense of flower and pod production. Seeds should be planted 1 to 2 inches (2.5 to 5 cm) deep. If soil is especially cold or wet, they may rot. 'Alaska' is especially tolerant of adverse soil conditions.

**Insects and Diseases:** The most common diseases are fusarium wilts, root rots, viruses, and powdery mildew. There is some varietal resistance. Pea aphids and pea weevils are the most common insects.

**Food Uses:** Cook green-shell peas fresh, canned, or frozen. When dry, you can harvest peas and store them for later use. The seeds may be either smooth or wrinkled when dry. Wrinkled-seeded peas (which are most of the varieties available) produce sweet, tender, fresh peas, but they taste overly sweet even when dry and may taste odd in soups. Smooth-seeded varieties to grow for dry peas have a starchy flavor that tastes good in soups, but if you harvest them early for fresh peas their flavor is mediocre. They are seldom available to home gardeners, except for 'Alaska,' a cold-hardy variety grown for early fresh peas.

Edible-podded peas, also called snow peas or sugar peas, lack the fibrous pod wall of other peas. The young pods have a crisp texture and sweet, delicate flavor. Use them fresh or frozen, steamed or stir-fried, alone or with chicken or beef. They have appeared in U.S. supermarkets quite recently, fresh and frozen. They are expensive and the texture is often flaccid. Fresh pods have often been stored for a long time. It pays to grow them yourself!

**Landscape Uses:** The varieties that are most attractive and most useful in the landscape are tall, vining ones. With some help, they can climb fences or house walls on net or wire. Trained on string or trellises, they make effective early-season screens and garden space dividers. Grow them alone or mixed with sweet peas for showier flowers. (Sweet peas are not edible.) Peas can be a backdrop for tall, mild-colored flowers that bloom in late spring or early summer such as iris, lily, leek, delphinium, bleeding heart, astilbe, and rose.

If you plant dwarf peas, surround them with hedges of erect, compact plants such as lavender, sage, and anise hyssop.

If you grow peas for their dry seeds, put them in the back of the garden, where they can wither without being conspicuous.

**Varieties:** 'Mammoth Melting Sugar' is a large-leaved, pink-flowered, edible-podded vine 4 feet (1.2 m) tall. 'Alderman' or 'Dark-colored Telephone,' an attractive variety for fresh shelling peas, has white flowers and grows up to 6 feet (1.8 m) tall. 'Sugar Snap,' a new white-flowered variety up to 8 feet (2.4 m) tall, produces both edible pods and good-tasting peas.

Some small varieties bear early and are quite disease-resistant. Good, small shelling varieties 1½ to 3 feet (45 to 90 cm) tall include vigorous 'Green Arrow,' heat-tolerant 'Wando,' and wilt-resistant 'Freezonian' and 'Progress No. 9.' White-flowered 'Oregon Sugar Pod' and lavender-flowered 'Dwarf Gray Sugar' are good small-vined varieties with edible pods.

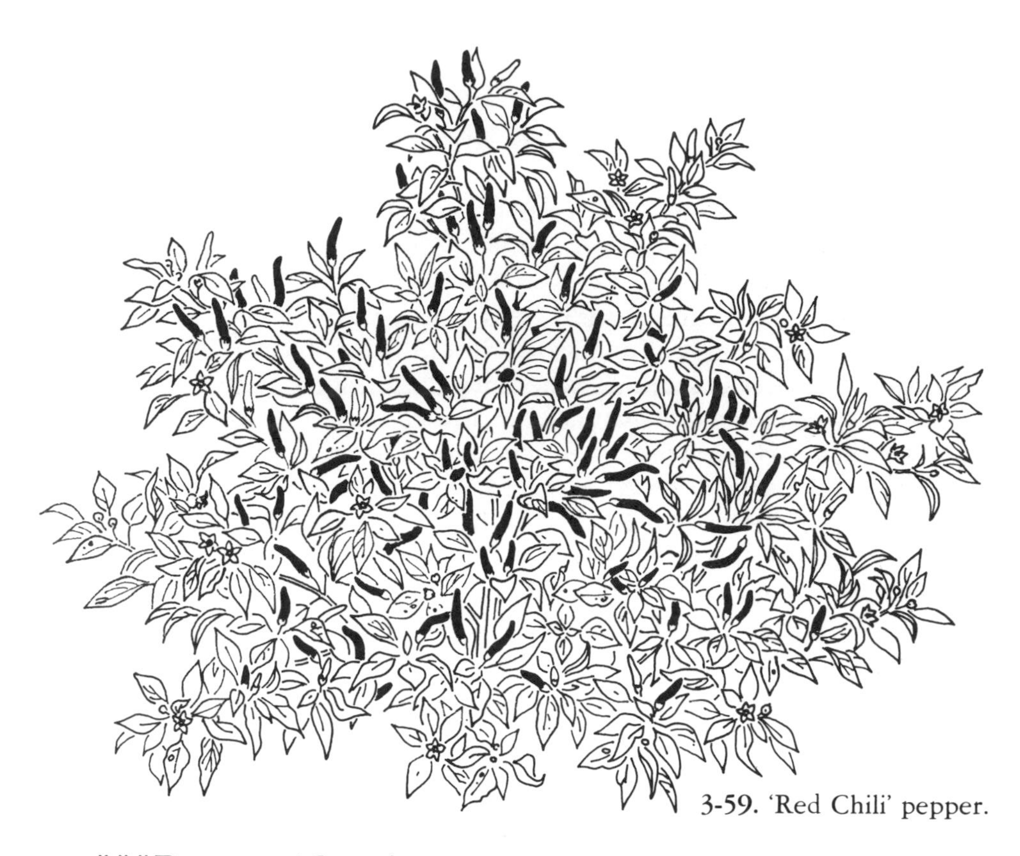

3-59. 'Red Chili' pepper.

## ***Pepper (*Capsicum annuum*)
tender annual

**Ornamental Characteristics:** There are hundreds of different pepper varieties and much variation in their characteristics. They range from fair to excellent ornamentals. Plants may be upright, or low and spreading. Leaves range from broad oval and blistered to narrow, small, and flat. (See figure 3-59.) They are clear, glossy green, and darker in some varieties than in others. Star-shaped white flowers may bloom briefly before the fruits begin to ripen, or bloom continuously even as fruits turn red, giving the plants a colorful, abundant air. Pepper fruits are green or yellow at first, turning red or orange as they ripen. Their colors are bright and cheerful against the foliage. Fruits may be upright or hanging, flame shapes, perfect spheres like Christmas balls, large, blocky "bells," or wrinkled cones.

The most attractive peppers have spreading growth, small leaves, continual bloom, and small fruits that turn color early in the season.

**Ornamental Drawbacks:** Peppers have shallow roots and brittle stems. Plant breakage is a big problem. Staking peppers or planting them in groups helps prevent it. Breakage may be greater when moisture and soil nutrients are plentiful, but these conditions also increase plant vigor and yield.



Many erect pepper varieties stay stiffly upright most of the summer. Then, as fruit turns color and plants look their best, large branches crack off or entire plants keel over and lie on their sides in the garden. They are seldom entirely uprooted. They continue to ripen fruit, but their effect as ornamentals is lost. The advantage of low-growing peppers (or those that collapse early in the season!) is that they are already on the ground and tend to grow in well-shaped mounds. Many hold their fruit well off the ground.

Some varieties have more fruit than foliage, and they look awkward, unnatural, and weighed down. Others have relatively few fruits, but the fruit size is huge in proportion to plant size.

Many bell peppers have large, thin, slightly blistered leaves that may curl up at the edges and are somewhat unattractive.

**Adaptability:** Peppers grow best in full sun. In partial sun they have fewer leaves and fruits and longer stems and often a distinctive, attenuated grace.

They produce fruit in long, cool summers, but do best in warm summers. An early start indoors is necessary for most climates. You can sow seeds in April and move plants outside in warm weather.

**Insects and Diseases:** Most pepper plants are very healthy, but there are many possible problems. Potential insects include leaf miners, aphids, flea beetles, cutworms, hornworms, pepper maggots, and pepper weevils.

Major diseases include damping off (seedlings), several leaf spots, root rot, ripe fruit rot, anthracnose, and an assortment of mosaic viruses. Peppers have two physiological difficulties as well—sun-scalded fruit and blossom drop.

**Food Uses:** Eat sweet and hot peppers fresh, frozen, canned, dried, or pickled. You can use certain varieties to make chili powder, cayenne, and paprika. As mild peppers ripen, they turn color and change from bland-tasting and juicy to sweeter and drier. Hot peppers become steadily hotter as they ripen.

**Landscape Uses:** Good pepper varieties are beautiful almost anywhere in the garden—in beds, in flower borders, alone or in masses, in rows as edging plants, in pots and in window boxes. They have a long season of attractiveness and harvest. In late summer, the glowing fruit colors combine well with yellow, orange, and red flowers. Foliage looks best with other bright greens, such as celery and parsley. Growing peppers close together in a bed rimmed by erect, bushy plants helps hold the peppers up; you can also stake them or grow them along a wall or fence.

**Varieties:** Peppers are divided into groups to reduce the confusion of so many varieties. The groups I am using are chili (Tabasco), cayenne, cherry, bell, and conical peppers.

Chili peppers are used in hot sauce, Tabasco sauce, curries, and pickles. 'Red Chili' and 'Fiery Festival' are beautiful and both are fiery hot. 'Red Chili' has small, narrow, bright green leaves; graceful, spreading growth 2 feet tall and 2 feet wide (60 by 60 cm); and a large number of thin, upright fruits that turn in early fall from pale, bright green to flaming red. 'Fiery Festival' is a compact, erect bush 1 foot tall and 2 feet wide (30 by 60 cm.) with small, dark green leaves; continually blooming flowers; and numerous, small, curved fruits that turn from yellow to orange and finally bright red. In autumn the plants look as though they are literally going up in flames. And they do not fall over.

Cayenne peppers are used in chili and cayenne powders, and range from slightly to breathtakingly hot. They are an unrewarding group visually. Large fruits of 'Anaheim' or 'New Mexico Chili' are used green for chili rellenos and green chili sauce. They are good-sized plants—2 to 2½ feet (60 to 75 cm) tall and 2 to 3 feet (60 to 90 cm) wide—with big, sparse, droopy leaves and long, wrinkled fruits. They are top-heavy and often fall over. The best-looking varieties are narrow-fruited ones such as 'Long Thin Cayenne' and Japanese 'Fushimi Long Green.' Both collapse early in summer and grow in low, attractive mounds. Fruits of 'Fushimi Long Green' are pale green in contrast to darker foliage. Surprisingly they taste mild rather than hot.

Cherry peppers may be hot or sweet. The hot varieties we grew in our trial gardens looked great until they collapsed in late summer. 'Cherry Sweet' fell over early and spread along the ground. It was good-looking and somewhat smaller than a cayenne plant. Its small, spherical fruits are often pickled.

Cherry pepper is *not* the same thing as *Solanum pseudocapsicum,* Christmas or Jerusalem cherry, which is a bushy, pepperlike plant with round, nonpungent, poisonous fruits. If you are in any doubt about whether a plant is *Capsicum* or *Solanum,* don't eat it.

Bell peppers have thick-walled, blocky fruits that are mild and juicy and eaten cooked or raw in salads. 'Keystone Resistant Giant' is exceptionally erect and vigorous, though its good-tasting fruits are partly hidden under its dense foliage. 'Atlas' is quite unattractive. Its huge fruits occupy the center of each plant in a solid, awkward-looking clump. Most standard bells, including widely available 'Yolo Wonder' strains, are intermediate in appearance. They break and fall less than many peppers, but when they do, they look very awkward. They are 1½ to 2 feet (45 to 60 cm) tall and 2 to 2½ feet (60 to 75 cm) wide. Leaves are large, thin, and often droopy. The chunky fruits are not gracefully shaped, but look pretty when they turn red.

Fruits of orange or golden bell varieties turn from green to rich, glowing orange, and do not become red. They taste excellent, especially when they are orange, but most fall over in late summer. The best is 'Golden Calwonder'; its fruits turn color later than do other varieties, but plants are unusually erect.



Conical bells look better than other bells. The slender bell-shaped fruits taper at the tip. They are early, thin-walled, and slightly drier than standard bells. Foliage is relatively small. The finest and earliest variety, Japanese 'Early Bountiful,' keeps flowering as its fruits turn deep red and spreads gracefully along the ground. 'New Ace' is quite erect. Its fruits turn from green to orange to red (flashy!) as it continues to flower. 'Shepherd' and 'Canape' are also attractive.

Conical peppers include both pimento and banana peppers. Pimento fruits are mild and dry-tasting, broadly conical, and thick-walled. They turn from green to red. When red, they are canned, stuffed into green olives, or ground into paprika. They can also be used in the same ways as bell peppers. They are only fair ornamentals. 'Pimento L' looks slightly better than do 'Pimento Select' and brown-fruited, weak-stemmed 'Sweet Chocolate.'

Banana peppers include several good-looking varieties. Fruits are thin-walled, but mild ones can be used as bell peppers are. They are slender, hanging cones, pale and shiny yellow at first, turning to deep yellow, orange, and red as they ripen. 'Early Sweet Banana' and 'Hot Hungarian Yellow Wax' are first to turn color; 'Early Sweet Banana' sometimes looks overburdened by long fruits. 'Sweet Banana' fruits have the best flavor. 'Petite Yellow Sweet' with its delicate fruits is the most attractive. It needs help to stay erect, as do they all. They are as tall as they are wide, 2 feet by 2 feet (60 by 60 cm), and their heavy crops often make them collapse.

## **Poppy, Shirley/Field Poppy (*Papaver rhoeas*)
hardy annual

**Ornamental Characteristics:** Rosettes of deeply cut, gray-green Shirley poppy leaves are attractive even before the flowers bloom. The flower buds, drooping teardrop shapes covered with hairs, rise from the rosettes on long stems and turn upward as they open. The flowers have thin, papery petals—white, pink, red, or scarlet—and bushy masses of stamens. As plants bloom, they branch and become broader. The flowers are followed by round "shaker" seed capsules, each with a lid on top that lifts when the seeds are ripe.

**Ornamental Drawbacks:** As they increase in numbers, the long, thin flower stems make the plants messy-looking. Shirley poppy plants in full bloom often seem to be at least 80 percent stems. The delicate flower petals are short-lived; in windy locations, flowers are blown apart and seldom last more than a day. Trimming off old flowers frequently keeps the plants blooming. Cool weather and plenty of growing space also prolong bloom. Plants are not attractive once they are left to mature seed.

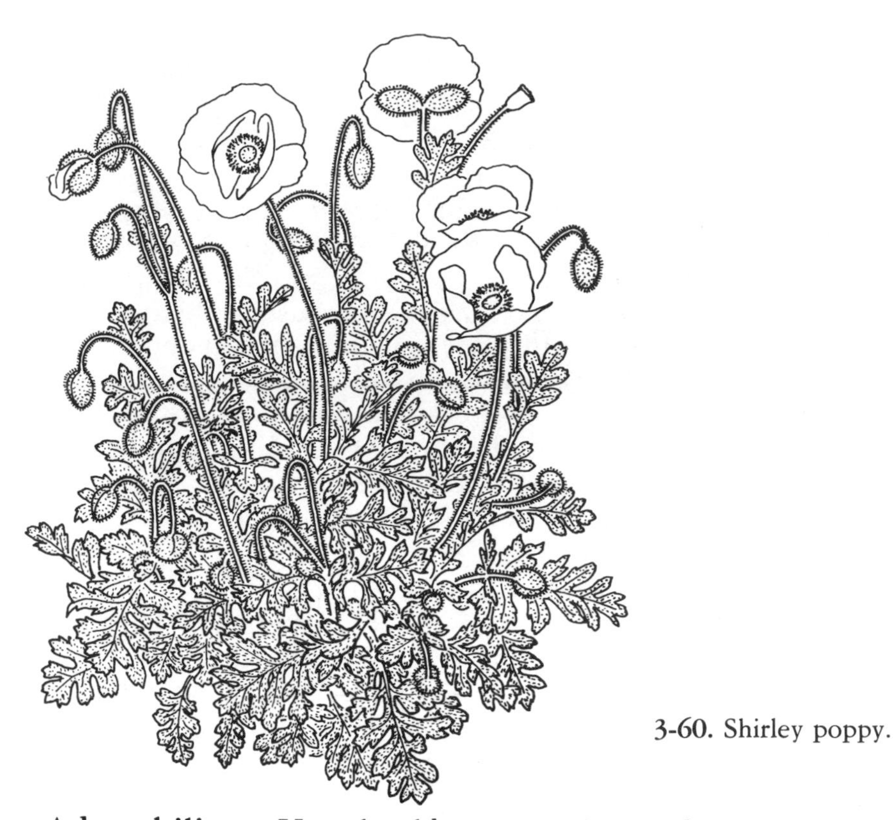

3-60. Shirley poppy.

**Adaptability:** You should sow poppies outdoors in place; the fragile roots cannot easily be transplanted. The tiny seeds germinate best in cool, loose-textured soil. In fall or early spring, you can scatter them and press them gently into the soil surface. Germination is often spotty.
**Insects and Diseases:** Poppies are susceptible to aphids, cutworms, and bacterial blight.
**Food Uses:** If you move poppy seed capsules back and forth and hear seeds rattling around inside, they are ripe. They should be eaten only when ripe, and used in moderation; large quantities may cause gastrointestinal upset. Shirley poppy seeds have been used in Europe for seasoning and as a source of bland cooking oil. When baked, they have a crunchy texture and warm, nutty flavor. They can be added to cakes, breads, rolls, cookies, salad dressing, rice, noodles, curries, and vegetables.
**Landscape Uses:** Poppies are at their best in spring and early summer. They can be planted near late spring-flowering bulbs or sown with field wildflowers and mild-mannered grasses to form a "wild lawn" for a front yard, sunny hill, or orchard. You might also plant them in masses and drifts near other plants with heavier foliage and greater staying power than theirs.



They look good near other gray-green leaves—sage or kale.
**Species and Varieties:** Poppy seed sold in grocery stores comes from the opium poppy (*P. somniferum*). It is imported from Europe and sterilized so it will not grow. Opium is made from the sap of seed

3-61. Fall radish.

capsules, but ripe seeds are safe to eat. Federal law prohibits growing opium poppy anywhere in the United States. Seed labeled 'Hungarian Blue-seeded' is liable to be opium poppy; it is a pretty plant, with large, lavender, tuliplike flowers.

Fortunately, another species, Shirley poppy (*P. rhoeas*) is legal and widely available. We grew the variety 'Shirley Mixed Single.' It grows up to 3 feet tall and 3½ feet wide (90 by 105 cm), with toothed, lobed, jagged-looking leaves, 1-foot (30-cm) flower stems, and 4-inch (10-cm), somewhat double flowers of red, scarlet, pink, and white. Seed capsules are quite small—about ½ inch (1.3 cm) across—compared to 1½ inches (3.8 cm) for opium poppy, but they contain plenty of dark brown seeds.

Seed of other *Papaver* species may sometimes be sold as Shirley poppy. Stokes and Burpee are two reliable sources of 'Shirley Mixed Single.' I know of no reliable source of 'Shirley Mixed Double' and do not recommend its use as a food plant. 'Flanders Field' is a striking variety with black-spotted red petals; it is often sold as *P. rhoeas,* but there is some question whether it actually belongs to this species. No other *Papaver* species are definitely known to be edible and some may be harmful.

## **Radish, Fall/Daikon (*Raphanus sativus*)
hardy biennial grown as an annual



**Ornamental Characteristics:** Large-rooted fall radishes are very different in appearance and food uses from the more commonly grown spring radishes with their small roots and plain, bristly, unat-

tractive leaves. Spring radish is listed under "Less Attractive Vegetables and Herbs" at the end of this chapter. The best fall radish varieties grow in broad rosettes; they look like large, bold-leaved ferns. The long, blue-green leaves are divided into many overlapping, round, scallop-edged leaflets lined up in pairs along each leaf stalk.
**Ornamental Drawbacks:** Some varieties have undistinguished leaves that grow in upright bunches. Even varieties with broad rosettes need plenty of space to grow to full potential; crowded plants hold their leaves high in the air and the rosette form is not evident. You can sow seed thickly, and thin as plants become crowded.
**Adaptability:** The ideal time to sow fall radishes is mid-July through early August. They may bolt if sown in spring; cold weather and long days induce them to flower. Seed germination is fast and even. An even water supply is needed because plants grow quickly. Roots become fairly large in two to three months. Unless the soil is very wet, you can store them in the garden during cold weather, although tops are killed back by severe frosts. In mild climates, radishes that are not harvested bloom the next spring.
**Insects and Diseases:** Cabbage maggots, flea beetles, slugs, aphids, and club root are possible problems.
**Food Uses:** In the Orient, radishes are grated in soy sauce as a dip for various foods. They are also steamed, boiled, stir-fried, pickled, brined, or dried. Root shapes range from long and straight to globular and enormous; external colors include pink, white, purple, and black, always with white flesh inside. Flesh texture of radishes is usually firm and solid; flavors range from mild to very hot.

Further uses of radishes include boiling the leaves as greens, pickling the seed pods, or brining them as caper substitutes. You might also add the peppery seeds to salads.
**Landscape Uses:** Grow fall radishes in rows or beds, near blue-green foliage of savoy cabbage, flower kale, fennel, leek, and gray-green sage, dusty miller, and wormwood. Rose-colored, pink, and white flowers—cleome, martynia, chrysanthemum, zinnia, and aster varieties—complement radish foliage.
**Varieties:** 'Sakurajima Mammoth' and 'All Seasons White' are excellent ornamentals. 'Sakurajima Mammoth' forms dense rosettes 1 foot tall and 2½ feet wide (30 by 75 cm), with beautiful blue-gray leaves; the round, white, pungent roots can grow up to 50 pounds (23 kg) if they have five months of frost-free weather. 'All Seasons White,' not a true fall radish, can be planted in either spring or fall. Its foliage, similar to that of 'Sakurajima Mammoth,' is almost as attractive. It is a slightly smaller plant, with long, white, pungent roots.



'Winter King Miyako' and 'White Chinese' or 'Celestial' have attractive, bright green leaves. Both varieties are taller and narrower

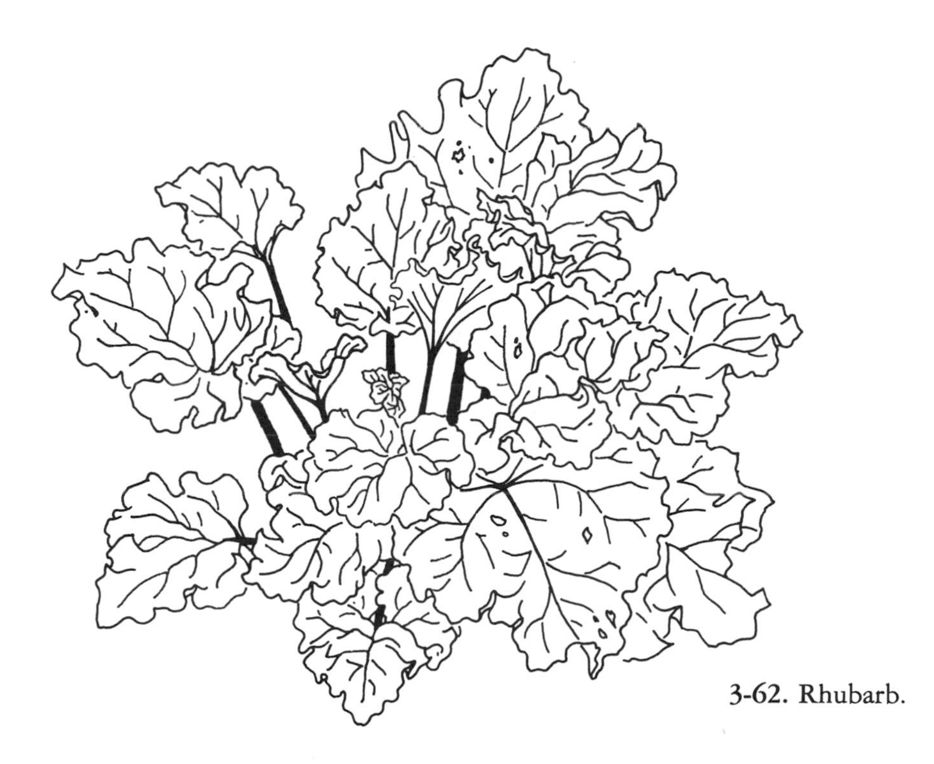

3-62. Rhubarb.

than 'Sakurajima Mammoth'—1½ feet tall and 2 feet wide (45 by 60 cm). Their leaves are more erect and do not form distinct rosettes, although 'Winter King Miyako' comes close.

'Round Black Spanish,' 'China Rose Winter,' and 'Summer Cross Hybrid,' an all-seasons type, are less attractive than the varieties described above.

Rat-tailed radish (*R. sativus* 'Caudatus') has long, twisted, purple seed pods that are harvested immature and eaten raw or pickled.

## **Rhubarb (*Rheum rhabarbarum*)
hardy perennial

**Ornamental Characteristics:** New rhubarb shoots pushing up through the soil in early spring are stocky and droll-looking. Their leaf stalks are green, pink, or red. The dark green leaves unfold and become large and broad, with blistered surfaces and wavy, scalloped edges. (See figure 3-62.) They grow in vigorous rosettes up to 4 feet (1.2 m) across. Plants may become 6 feet (1.8 m) tall in flower; erect flower panicles are greenish white and feathery. Rhubarb has a lush, tropical look, especially when it is blooming.
**Ornamental Drawbacks:** Rhubarb seed stalks are conspicuous and weedy-looking.
**Adaptability:** Leaves die at about 27° F. (−3°C.), but roots are thoroughly hardy; rhubarb is a Siberian native. It grows best where winters are cold and summers mild and does not flourish in the south-

eastern United States. Its fleshy roots require a lot of nutrients and water; it should be watered throughout the growing season.

Flowering may make rhubarb stalks bitter. Pick off flowers until you are finished harvesting the stalks. Then allow the flowers to bloom and remove them as they go to seed.

Rhubarb becomes less vigorous after four to twelve years. Stalks become short and thin, even early in the growing season. At this time, you can divide old plant crowns and plant the pieces in new soil during fall, or, preferably, spring. (Rhubarb is not generally started from seed, because seedlings are variable and seldom resemble their parents.) Each piece of crown needs at least one strong bud on it; plant pieces several inches deep, in a place where their roots will not be disturbed. Do not harvest them during the first year of growth. In the second year, a four-week harvest is best. During subsequent years, harvest can last six to eight weeks.

**Insects and Diseases:** Rhubarb is usually quite healthy. Slugs, snails, rhubarb curculio beetles, and crown rot (foot rot) are possible problems.

**Food Uses:** Make rhubarb stalks into jams, jellies, preserves, pies, cookies, and sauces for ice cream or poultry. Cook them alone or combined with other fruits such as strawberries, pineapples, and oranges. Can or freeze them, if you wish.

Several old herbals mention rhubarb leaves as a potherb, but their use has resulted in fatal poisonings. Rhubarb leaves contain soluble oxalic acid. Sensitivity to oxalic acid is an individual matter, and what is okay for one person may make another one quite sick. The flowers are occasionally suggested as food, but they are poisonous to some people and not worth experimentation. Cooked rhubarb stalks, however, are safe for everyone. Although they contain relatively small amounts of oxalic acid, most of it is in insoluble form. Green stalks are as edible as any other kind; cold weather and varietal differences cause pink and red coloration.

**Landscape Uses:** Grow rhubarb plants in beds or borders, alone or in masses. Line them up as low hedges. Rhubarb in a big tub lends a strong exotic touch to a patio. Do *not* plant rhubarb any place where small children might eat the leaves.

Rhubarb makes a good background plant for big flowers of late spring and summer such as tulips, lilies, day lilies, bee balm, gladiolas, zinnias, and tall marigolds. Rhubarb can also back up colorful foliage plants such as 'Rhubarb' Swiss chard, 'Dark Opal' basil, and amaranthus varieties. If you plan to let rhubarb plants bloom, astilbe and ornamental grasses have flowers similar to theirs. Plant them where their emerging sprouts show up well in the spring, perhaps with crocus and grape hyacinth.



**Species and Varieties:** 'McDonald,' 'Ruby,' 'Valentine,' and others

3-63. *Rosa rugosa.*

are red-stalked varieties. 'McDonald' is especially vigorous. 'Victoria,' 'Linnaeus,' and 'Sutton' are sturdy varieties with stalks that are green or pink under garden conditions. Their stems are red only if plants are forced in the dark.

Reports differ on whether related species—*R. australe, R. nobile, R. palmatum, R. ribes,* and *R. tataricum*—are edible or not. It is not a good idea to experiment with any of them, since they may contain purgative and/or poisonous properties.

## ***Rose (many *Rosa* species)
hardy perennials

**Ornamental Characteristics:** For many centuries, gardeners have loved the beautiful, fragrant flowers of roses. There is a wide range of flower colors, shapes, sizes, and scents. Flowers may bloom all at once, usually in early summer, or throughout summer and early fall. They are followed by oval, red fruits or "hips." The dark green, ovate leaves with toothed margins may be large or small, shiny or dull-surfaced. Stems usually bear thorns but are occasionally thornless. Growth habit may be climbing or upright, narrow or spreading. Plants grow from 3 inches to 8 feet tall (8 cm to 2.4 m).
**Ornamental Drawbacks:** Large-flowered, upright roses often have stiff, narrow growth habits and look more like statues than plants.

Many roses are plagued by disfiguring insects and diseases; growing healthy plants requires constant vigilance. In most climates, roses shed their leaves in fall and spend the winter as prickly sticks.

**Adaptability:** Roses need room to grow. Hybrid roses (those usually seen in gardens) need very fertile soil and plenty of water, but most species roses are less demanding. Roses can be propagated by cuttings, budding, or grafting. Occasionally, species roses are grown from seed. Although they can be planted in fall or spring, roses are most often available in spring. The shallow roots benefit from mulching, and plants should be pruned in spring and summer to remove dead, infected, or twiggy branches and to encourage vigorous, well-shaped growth with plenty of flowers. Pruning systems vary according to gardener and rose type. In general, species roses require less pruning than do hybrids.

Many roses have unreliable cold-hardiness below 15° F. (−9° C.). Heaping straw or earth around them helps reduce temperature fluctuations. If roses are killed back to the ground, new canes usually appear in spring.

**Insects and Diseases:** Some gardeners spray or dust their roses each week. They enjoy the flowers so much that they find this procedure worthwhile. If you plan to eat your roses, make sure sprays or dusts are safe for food plant use.

Possible insect problems include aphids, leafhoppers, thrips, Japanese beetles, spider mites, and sawflies. Diseases include black spot, mildew, canker, and rust. So far, hybrid roses have been bred for beautiful flowers more than for disease resistance; species roses generally have fewer diseases.

**Food Uses:** Fresh rose petals lend a fragrant, exotic flavor to salads, omelets, pancakes, juices, wines, liqueurs, teas, and desserts. You can dry them or grind them up for use as seasoning, or preserve them in jelly, jam, or conserve. They add spice to sugar, butter, or vinegar. Their flavor and scent are intensified in rose syrup and rose water, which are used frequently in Middle Eastern and Indian cooking. For eating purposes pick flowers when they are still fresh and trim the bitter bases from the petals.

Eat rose hips raw, or make them into jelly, jam, tarts, sauces, soups, tea, wine, and juice. They have an acid, fruity flavor. When fresh, they contain large amounts of vitamin C. Much of it is lost when the hips are dried, but it can be preserved if they are frozen. Only the fruit walls of rose hips are eaten; the seeds inside are discarded. A few rose species have thorny-skinned hips that cannot be eaten at all.



**Landscape Uses:** Climbing roses, tied to fences and trellises, make a good green background for a wide variety of other plants, and when in bloom—usually during early summer—they are very showy. Foreground plants of contrasting flower colors are striking. Bushy roses—

# Vegetable and Herb Flowers

**C-1.** Summer squash.

**C-2.** 'Red' okra.

C-3. Amaranthus love-lies-bleeding.

C-4. Leeks in bloom. (Roy and Marsha Rathja's garden, Corvallis, Oregon.)

C-5. Anise hyssop.

C-6. 'Chabaud's Giant Improved, carnation.

## Leaf Colors of Vegetables and Herbs

**C-8.** 'Coral Queen' flower kale.

**C-7.** Zucchini squash.

**C-9.** Iceberg lettuce.

**C-10.** 'Curlina' parsley.

# Leaf Textures of Vegetables and Herbs

C-11. 'Early Jersey Wakefield' cabbage.

C-12. Creeping thyme.

C-13. 'Red Chili' pepper.

C-14. 'Royal Chantenay' carrot.

C-15. 'Delicata' winter squash.

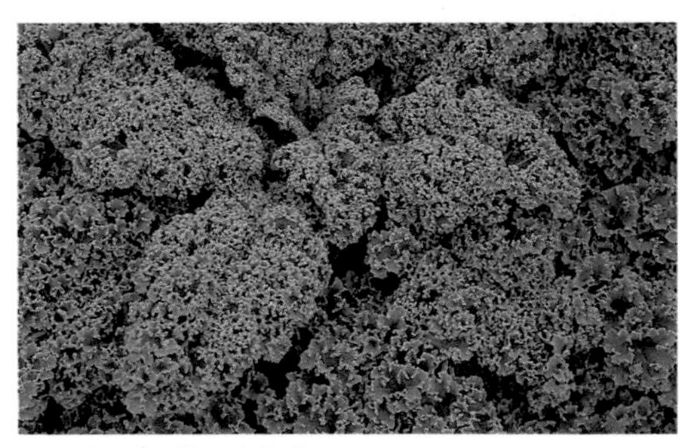

C-16. 'Petibor' kale.

# Spring Plant Combinations

**C-17.** Borage and lovage.

**C-18.** 'Concord' grape and 'Flanders Field Single' poppy. (Photo by George Gessert, Jr.)

## Late Summer Plant Combinations

C-19. Corn and sunflowers.

C-20. 'Golden Delicious' winter squash and calliopsis.

C-21. Clockwise from lower left corner: heliotrope, verbena, lavender, geranium, sweet pepper, flower kale, parsley. (Anita and Jim Green's garden, Corvallis, Oregon.)

C-22. Pool with arrowhead, water chestnut, water lily, and parrot's feather, rimmed by winter squash. (Designed by George Gessert, Jr.)

C-23. Clockwise from lower left: geranium, verbena, sweet pepper, Chinese chives, 'Burgundy Delight' corn, dahlia, marigold, flower kale, parsley, heliotrope. (Anita and Jim Green's garden, Corvallis, Oregon.)

C-24. Clockwise from lower left: 'State Fair Mixed' zinnia, 'Dwarf Blue Scotch' kale, 'Diablo' cosmos, mixed corn cultivars, yellow zinnia, Florence fennel, 'Lulu' marigold, 'Petite Yellow Sweet' pepper. (Author's garden at Oregon State University Vegetable Research Farm, Corvallis, Oregon.)

species roses, "shrub roses," and cluster-flowered roses—look good in clumps and hedges. You can use taller ones in the background of beds and borders. Miniature roses look best in containers or rock gardens; their intensely dwarf habit gives an overprecise air to beds in which they are planted.

Although narrow, upright roses—the familiar hybrid tea roses and grandifloras—are difficult to integrate into the garden, they can be used in the background of a border. As much as possible, prune them into soft shapes to avoid a stiff, statuary effect. The flowers are so large, bright, and distinctively shaped that it is hard to combine them with other flowers. Instead, grow attractive foliage plants nearby, such as celery, parsnip, Italian parsley, zucchini squash, or a background of corn.

**Species and Varieties:** There are hundreds of beautiful hybrid roses, and you may already have favorites. Hybrid rose groups include climbers, large-flowered roses (hybrid tea and grandiflora), cluster-flowered roses (floribunda), polyantha roses (with clusters of small flowers), shrub roses, and miniature roses. All roses are edible, but the most fragrant, good-tasting flowers and the most succulent fruits are found among the species roses. As a bonus, species roses are easiest to maintain. Their major drawback is that they do not ordinarily bloom all summer, while many hybrids do.

*Rosa centifolia, R. damascena,* and *R. gallica* bear strong-scented flowers in early summer. *R. centifolia,* cabbage or Provence rose, reaches 6 feet (1.8 m) in height and has nodding, double pink flowers. *R. damascena,* damask rose, 5 feet (1.5 m) tall, bears clusters of double pink or red flowers. *R. gallica,* French rose, has intensely fragrant, single red flowers and grows to 4 feet (1.2 m). It is the rose most often used in cooking. One variety, 'Officinalis,' with double pink flowers, was so frequently grown for medicinal use during the Middle Ages that it became known as "apothecary rose." All three species are vigorous and long-lived.

Species with the best rose hips include *R. rugosa, R. villosa (= R. pomifera),* and *R. spinosissima.* Six-foot (1.8-m) rugosa rose or sea tomato (*R. rugosa,* shown in figure 3-63), has bright 1-inch (2.5-cm) fruits that are especially high in vitamin C. Its large, fragrant flowers bloom all summer; varieties may be single- or double-flowered, white or bright pink. Flowers are strong-tasting enough for food use. *R. rugosa* makes a good dense hedge plant. Its wrinkled, shiny leaves turn brilliant orange in the fall, and it has excellent cold-hardiness. *R. villosa,* apple rose, a densely branching bush of 6 feet (1.8 m), has single, pink flowers, and fruits up to 1½ inches (4 cm) long. *R. spinosissima,* burnet or Scotch rose, has small, shiny leaflets and prolific flowers, single or double, pink, white, or yellow. Its small purple-black fruits are not prominent, but taste unusually sweet. It grows in dense,

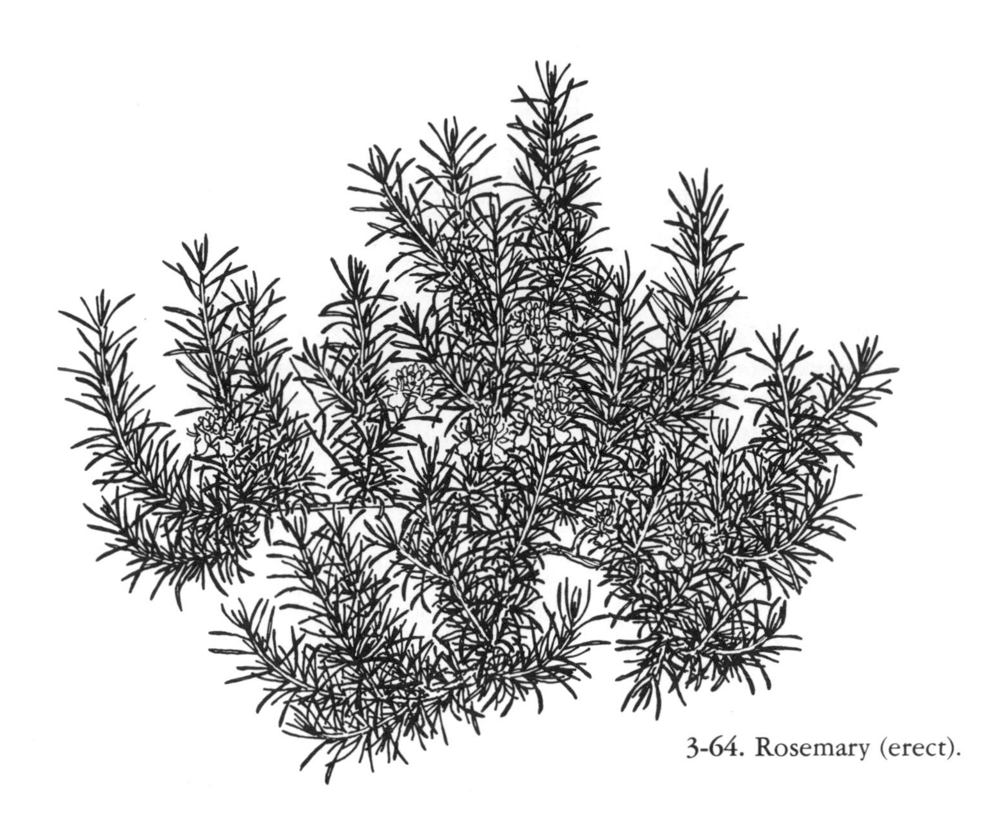

3-64. Rosemary (erect).

thorny, 3-foot (.9-m) mounds and does well in poor, dry, sandy soil; it may spread rampantly in rich garden soil. The botanical variety *altaica* has beautiful, single white flowers and good, vigorous growth, 4 to 6 feet (1.2 to 1.8 m) tall.

Other species sometimes used as food include *R. canina, R. cinnamomea, R. nutkana,* and *R. eglanteria.*

## ***Rosemary (*Rosmarinus officinalis*)
hardy perennial

**Ornamental Characteristics:** Rosemary plants have a fine, precise texture. Their narrow leaves, arranged neatly on the stems, resemble fir or spruce needles. They have a pinelike, warm and spicy scent. (See figure 3-64.) They are medium green and leathery on top, soft and gray-green underneath. Their growth habit is variable. The species grows in erect bushes, and there are prostrate forms as well. Plants bloom in late winter or early spring. The pale blue flowers resemble tiny, open-faced snapdragons and appear in leaf axils along the stems.
**Ornamental Drawbacks:** Rosemary grows slowly, about 6 inches (15 cm) a year in most conditions. Erect rosemary grown in less than full sun, or watered too frequently, develops long, rambling branches. Pinching growing tips helps keep plants bushy.
**Adaptability:** Rosemary grows wild along Mediterranean coasts and even in the Sahara desert. Where well-adapted, it may grow 6 feet (1.8 m) tall.

3-65. Prostrate rosemary, at bottom, trailing down a retaining wall in Santa Barbara, California.

Erect rosemary is hardy to at least 12°F. (−11° C.). I have lost it during milder winters when its roots stayed wet; it requires good drainage. Prostrate rosemary is less hardy. In many climates you must overwinter rosemary indoors. It is an attractive, temperamental house guest. It needs as much light as possible, cool temperatures, and high humidity. It will not tolerate overwatering and responds to very dry soil by dropping all its leaves and sometimes dying. White flies, spider mites, and aphids all attack it under indoor conditions.

Plants are propagated by cuttings and, less commonly, by layering and division. Seed germination is slow and sometimes nonexistent. Warm soil seems to improve it. Seedlings get off to a very slow start. Plants that were overwintered indoors can be moved outdoors in pots during the summer or planted in open soil.

**Insects and Diseases:** There are no problems outdoors unless plants bring their indoor pests along with them.

**Food Uses:** Rosemary leaves are excellent seasoning for vegetables, rice, stuffings, stews, fish, meat, and poultry. Use them fresh, frozen, or dried. Make them into rosemary salt, rosemary butter, or rosemary sugar. Rosemary tea is a pleasant, pine-scented herbal remedy for colds, indigestion, and headaches. Use rosemary flowers in conserve or for seasoning. Bees make excellent honey from the flowers.

**Landscape Uses:** In cold-winter climates, most rosemary plants available for landscape use (from nursery stock or homegrown seeds and cuttings) are small—1 foot by 1 foot (30 by 30 cm) or less. Older plants that are overwintered indoors may get leggy and need to be pruned back to a small size. Plan for small-scale use.

3-66. Sage. (From Mm. Vilmorin-Andrieux's *The Vegetable Garden.*)

Erect rosemary is an excellent low edging, alone or with dianthus, phacelia, dwarf phlox, lobelia, or blue-starred oxypetalum. It can be backed by larger plants such as sage and lavender. It can be planted in borders, beds, or rock gardens. It is a good pot or window box plant. It blends well with other plants of fine texture and subdued foliage color—gray-green or dull green—such as yarrow, summer savory, thyme, lamb's ears, and small-leaved succulents.

Prostrate rosemary works very well in window boxes and hanging baskets, in rock gardens, and along retaining walls, where it can trail downward. (See figure 3-65.) At nurseries it is sometimes possible to buy fairly large prostrate rosemary plants, 1 foot tall by 2 feet wide (30 by 60 cm). In mild-winter climates, prostrate rosemary can be used as an evergreen ground cover. Erect rosemary can be massed in beds or lined up in hedges. It will grow taller and more impressive each year.
**Varieties:** Some varieties have white or dark blue flowers, silver-striped or gold-striped leaves. Variety names are inconsistent.

## ***Sage (*Salvia officinalis*)
hardy perennial

**Ornamental Characteristics:** Sage is a dense, erect bush up to 2 feet tall and 3 feet wide (60 by 90 cm). The oblong, smooth-edged leaves are pale gray with closely veined, pebbly-looking surfaces. (See figure 3-66.) Flower spikes, which are usually dark lavender, bloom prolifically for several weeks in late spring; flowers may still appear intermittently for a month or two afterward.



Sage keeps its leaves during moderately cold winters. Older stems are woody, with tender new growth at the tips.

**Ornamental Drawbacks:** Lower stems on older plants may become bare of leaves. You can avoid this by annually cutting the stems back to several inches or by making cuttings to replace old plants every four or five years.
**Adaptability:** Sage grows fast and easily from cuttings or seeds. Germination is excellent, and vigorous seedlings grow big the first summer. Layering and division are occasional propagation methods. Sage thrives with little care in a wide range of conditions, as long as it has well-drained soil and full sun.
**Insects and Diseases:** Sage is very healthy.
**Food Uses:** You can use fresh, dried, or frozen sage leaves, and flower petals as well, to season sausages, poultry, pork, stuffings, pickles, cheeses, and sauces. Sage has an almost medicinal smell, and its flavor is strong and rather astringent, so use sparingly. Sage leaves often have been used for tea, alone or with lemon and honey, as a remedy for sore throats, colds, fevers, headaches, and indigestion.
**Landscape Uses:** Sage is sturdy and versatile. You may want to plant more of it than you can eat simply for the pleasure of growing such dependable plants in your garden. It forms a low, uniform hedge and looks good in mixed borders, alone or in masses. It is attractive in pots and window boxes. It is easy to combine sage with bold or fine-textured, small or fairly large plants, and it is a good, erect companion for plants that have trouble standing up straight.

Plant sage with gray-green plants such as rosemary, lavender, and eggplant, and with blue-green plants such as leek, bunching onion, and kale. Many of these foliage combinations are effective during summer, fall, and mild winters. Sage foliage is a good background for the blue and purple flowers of heliotrope, nierembergia, campanula, and lobelia, and the yellow and orange flowers of marigold, butterfly weed, and gold yarrow. Sage flowers combine well with blue or purple flowers.
**Species and Varieties:** Names for sage varieties are confused. Go by plant appearance (or, in catalogs, plant description).

Plain sage is usually listed with no variety name. 'Broad Leaf' is occasionally available. Sage may have white, pink, or blue flowers instead of dark lavender. There is a red-leaved form and a variegated form, with white-edged leaves that sometimes have purplish coloration as well. Variegated sage is often known as 'Tricolor,' but seed labeled 'Tricolor' sage may turn out to be Joseph sage (*S. viridis*).

Pineapple sage (*S. elegans*) has downy-textured, toothed leaves, and spikes of good-sized red flowers in late summer. It grows to 4 feet (1.2 m) and may become leggy. It cannot tolerate frost, but can be overwintered indoors; cuttings root easily. The pineapple-scented foliage can be added to fruits, desserts, and drinks.



Joseph sage (*S. viridis*) has small flowers in leaf axils, and showy

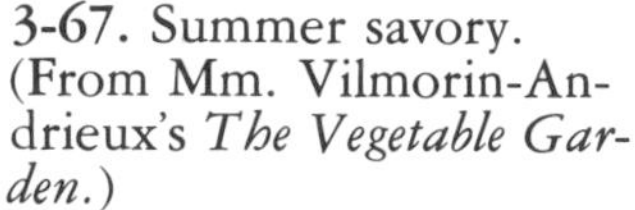
3-67. Summer savory. (From Mm. Vilmorin-Andrieux's *The Vegetable Garden.*)

pink, purple, or white bracts that cover the top foot (30 cm) of its many stems and stay bright-colored for six weeks or more. It is a large, dense-textured plant, up to 2 feet tall and 4 feet wide (.6 by 1.2 m), a fast-growing annual that may reseed itself. In Europe its leaves were once added to salads and cooked greens.

Clary sage (*S. sclarea*) is a self-sowing biennial or perennial with large, wooly leaves, conspicuous white or lavender bracts, and resin-scented blue or lavender flowers. The botanical variety *turkestaniana* has extra-large bracts and pink and white flowers. Young leaves can be fried, made into fritters, or added to omelets.

## ***Savory, Summer (*Satureja hortensis*)
half-hardy annual

**Ornamental Characteristics:** Summer savory has small, narrow, gray-green leaves borne opposite each other on purplish branches. These give the plants a neat, fine texture. Like a candelabrum, each plant has a straight main stem that branches halfway up. The branches are in opposite pairs with each branch at right angles to the one below it. This arrangement gives geometric regularity to the plants, especially when they are young. Plants grow up to 1½ feet tall and 1 foot wide (45 by 30 cm).

In late summer, as the plants begin to bear pink flower clusters along the stems, the leaves turn purplish and stems become wine red. The colors contrast subtly, and, from a distance, they look like a haze of warm, soft shades.

3-68. Winter savory. (From Mm. Vilmorin-Andrieux's *The Vegetable Garden.*)

**Ornamental Drawbacks:** Plants may fall over from top-heavy branching.
**Adaptability:** Sow seeds from early spring through early summer; they germinate easily. Plants should be thinned to only 6 inches (15 cm) apart so they can hold each other up. They are fairly drought-tolerant. They die when frosts begin or after flowering, whichever occurs first.
**Insects and Diseases:** Summer savory is free of pests.
**Food Uses:** Use summer savory fresh, dried, or frozen, to season fresh and dry peas and beans, poultry, meat, fish, sausage, dressings, eggs, soups, stews, and salads. Its flavor is warm and slightly peppery.
**Landscape Uses:** Plant summer savory in small clumps, big masses, or dense, low hedges. Savory can also grow in containers in protected places. Its neat leaves blend well with other leaves that are not overwhelmingly large or bold-textured. Blooming summer savory is a great contrast to bright green foliage such as that of celery, carrot, and parsley, or to bright yellow flowers. You might also grow it along with colors similar to its own, such as those of 'Dark Opal' basil and 'Molten Fire' amaranthus. Pink or lavender flowers, such as aster, cosmos, alyssum, and scabiosa varieties, are also good companions.
**Varieties:** None are available in seed catalogs.

## **Savory, Winter (*Satureja montana*)
hardy perennial

**Ornamental Characteristics:** Winter savory has narrow, dark green leaves and red-brown stems that become woody with age. The stems grow along the ground, but turn upward at the tips, where leaves are a lighter green. (See figure 3-68.) Each plant grows to be 6 to 9 inches (15 to 23 cm) tall and 1½ feet (45 cm) wide the first year from seed. During summer, fragrant white, pink, or lavender flowers bloom in the leaf axils; bees are attracted to them. Well-established

plants growing in full sun produce plenty of flowers that cover the upper stems.
**Ornamental Drawbacks:** Blooming is spotty in partial shade or during the first summer from seed. Frequent pruning keeps plants bushy for several years, but eventually they become sparsely leaved and need to be replaced.
**Adaptability:** Plants can be propagated from cuttings, layers, or division, or seed can be sown in spring or early summer. The seedlings grow slowly at first, so you may want to start them early indoors. Winter savory grows best in full sun, but does fairly well in partial shade. In cold-winter climates, well-drained soil helps plants develop good winter-hardiness. In many areas the leaves are evergreen.
**Insects and Diseases:** Plants are very healthy.
**Food Uses:** Winter savory is used the way summer savory is, but tastes stronger and more peppery. It is most useful in winter, when its leaves become milder and when summer savory is not available fresh.
**Landscape Uses:** Winter savory is often used as a low edging plant and does well in containers. It also blends well with fine- or medium-textured plants. It is such a dark green that it looks somber growing alone in large masses. It is best to plant it in small clumps with bright green plants nearby. The flowers go well with small spring bulbs and with other small flowers of blue, pink, lavender, purple, white, or yellow.
**Varieties:** 'Coerulea' flowers are dark lavender, a deeper shade than those of the species.

### **Sesame (*Sesamum indicum*)
tender annual

**Ornamental Characteristics:** Sesame's soft, hairy leaves are bright, light green. Some are ovate with slightly lobed margins, while others have three distinct lobes. The leaves grow in a precise pattern—four-square on one big, erect stem—but their fuzzy texture and variable shapes soften the impression of precision (see figure 3-69.) Sesame's height ranges from 1½ to 6 feet (45 cm to 1.5 m). The height may change with variety and environmental conditions.

Sesame begins blooming about two months from seed. Its good-sized flowers, borne in leaf axils up and down the stem, are shaped like those of penstemon—tubes open out into widespread, ruffle-edged petals. The tubes are soft pink and the petals are cream-colored. Angular seedpods, containing up to a hundred seeds each, split open when ripe.
**Ornamental Drawbacks:** Sesame is very attractive when healthy, but may quickly become disfigured by disease.
**Adaptability:** Sesame needs to be sown in warm soil, and although

3-69. Sesame.

it grows in most climates, it does best in the long, hot summers of the southern United States. It is very drought-resistant. Since sesame takes up to four months to produce ripe seed, some sources suggest starting it indoors up to two months early. Try this procedure only in short-summer climates where there is no alternative. Indoors, sesame may become infested with aphids or white flies, and plants started very early may flower indoors while too small to produce a good crop.

**Insects and Diseases:** Sesame is often very healthy, but one year in our trial gardens it was killed by a disease we were unable to identify.

**Food Uses:** Sesame seeds have a multitude of uses. Use them raw or toasted in baked goods, granola, candy, and stir-fries. Sprinkle them on fish, chicken, vegetables, soups, and desserts. In the Near East they are ground to make tahini (sesame paste). Harvest sesame when the first seeds at the plant base are ripe-colored (cream in most cases) and the pods at the top of the plant are green but well filled with seeds. Harvest the whole plant and hang it upside down in a paper bag in a warm, dry place. As pods open, seeds fall into the bag.

Young sesame leaves are edible too. They can be eaten raw, used sparingly in salads, or cooked as potherbs or in stir-fries.

**Landscape Uses:** You can grow sesame as a low hedge or edging plant, alone or with other plants of similar narrow, upright habit, such as anise hyssop. Sesame can support plants of uncertain erectness. It adapts itself to pots and window boxes, but clumps of sesame look good growing in beds too. In cool summer climates, sesame grows best along a south wall or in pots on a hot patio.



Sesame foliage blends well with light green, soft-textured grape leaves, gray-green lagenaria, and bright green parsley. Pale or yellow-green lettuce varieties are a shiny contrast to its matte surface. Sesame

**3-70.** Soybean. (From Mm. Vilmorin-Andrieux's *The Vegetable Garden.*)

flowers can be emphasized by similar-colored flowers nearby, such as unicorn plant, nicotiana, and pink and cream-colored snapdragons and petunias.

**Varieties:** Information on sesame varieties is minimal. Most seed catalogs list only the species. There are strains with gray, tan, cream-colored, dark red, and black seeds. The most common seeds are cream-colored when raw, golden brown when toasted. You can try growing seeds from Oriental groceries as well as those from catalogs. Some forms have dark pink flowers.

## **Soybean (*Glycine max*)

tender annual

**Ornamental Characteristics:** Soybeans have heart-shaped, prominently veined leaflets that grow in groups of three. Older leaves are fuzzy and gray-green. In bright contrast, new growth at stem tips is often yellow-green. (See figure 3-70.) Some varieties have erect, well-branched stems and form dense, compact bushes. Others are loose and spreading, but very attractive in their intense vigor. Varieties range in size from 1½ feet tall and 1½ feet wide (45 by 45 cm) to 4 feet tall and 4½ feet wide (1.2 m by 1.4 m).



**Ornamental Drawbacks:** Some varieties are sparsely leaved. Some have stems that fall to the ground late in the growing season. These need to be staked to remain erect. Small greenish white or lavender flowers and fuzzy green pods are fairly inconspicuous. By the time the beans ripen for dry seed use, the plants that bear them are dying and are no longer attractive.

**Adaptability:** For a succession of crops, plant varieties with different maturity dates. Dates range from 90 to 115 days from planting to dry-seed crop, but allow extra time during cool summers. Soybeans reach the green-shell stage about one month before they are fully ripe for drying. Seeds need warm soil to germinate and plants grow best in warm weather. They fix nitrogen in nodules on their roots. Adding undiseased soybean plants to the compost pile at the end of the season, therefore, or turning them into the soil, will enrich your garden.

**Insects and Diseases:** Soybeans are often healthy, but may be attacked by several bacterial, fungal, and viral diseases. Virus-free seed and resistant varieties are good protection against some of them. Possible insect pests include green clover worms, army worms, thistle caterpillars, leafhoppers, grasshoppers, mites, cucumber beetles, and blister beetles.

**Food Uses:** In the United States, most unexported soybeans are fed to livestock. However, in Asia, for many centuries millions of people have lived on a basic diet of rice and soy. The long list of products made from soybeans includes soy milk, tofu, tempeh, miso, soy flour, soy grits, soy sauce, soy oil, soybean paste, and bean sauces. Some of these foods are becoming popular in the United States.

Green-shell soybeans are not for sale here except in Oriental markets of large cities, but you can grow your own. Harvest them as the pods swell. Then boil or steam them, pods and all, and squeeze the seeds out after cooking. Eat them alone or with soy sauce, butter, or cheese. Or you might add them to stir-fries, soups, salads, and vegetable casseroles. They taste warm and nutty, and, when very small, they are as sweet as young corn. Later they become starchier, but taste good until just before they ripen into final hardness.

Harvest mature, dry soybeans as the pods turn brown. You can store the beans in or out of the pods, as long as they are completely dry. Soak them during winter and cook them in casseroles. Cooked soybeans can be pickled, or roasted and salted for snacks. Dry soybeans can be soaked, made into soy milk, and then made into tofu. Add soy sprouts to soups, stir-fries, omelets, and salads.

**Landscape Uses:** Use erect soybean varieties as hedges, and wobbly ones massed together in clumps. If you are growing soybeans for dry seeds, make sure the plants will be inconspicuous as they turn brown.

Black-eyed Susans and coneflowers, with their hairy gray-green foliage and gold and brown daisylike flowers, are ideal companions for soybeans. Other gray-green plants such as sage, Joseph sage, artemisia, catnip, borage, and nicotiana can also be grown near soybeans.



**Varieties:** Hundreds of soybean varieties have been developed in the Orient. An increasing number are available in the West as soybeans gain popularity in home gardens. Varieties vary in maturity dates, plant growth habits, and seed types. Seeds may be large or small, solid-colored or spotted, black, yellow, brown, gray, or white. Most small-seeded ("agricultural") varieties are grown for oil and livestock food, and large-seeded varieties for eating and tofu-making. Agricultural varieties are sold for food use in many grocery stores, but take a long time to cook and taste heavy instead of mild and nutlike. If you grow your own soybeans, you know what you are getting.

The varieties we grew are for eating rather than oilseed use. The earliest were 'Takii's Extra Early' and 'Fiskeby V.' Both were compact plants; 'Takii's Extra Early' was more erect than was 'Fiskeby V,' but 'Fiskeby V' had prolific yield. Its seeds are smaller than those of the other varieties and it seems well adapted to cool summers.

Several weeks after these varieties, 'Frostbeater' and 'Envy' began to bear. 'Frostbeater' germinated poorly but bore sweet, delicious seeds; 'Envy' had unusual purplish leaf coloration. After a bad start with poor germination, 'Prize' grew to be a vigorous plant with widespread branches and attractive leaves. 'Pickett' was the last to bear and was also the largest, most attractive variety. Its bright green leaves were very fuzzy and plants stayed erect despite their 4-foot (1.2-m) height.

## *Spinach, New Zealand (*Tetragonia tetragonioides*)
half-hardy annual

**Ornamental Characteristics:** New Zealand spinach has glossy, succulent, triangular leaves of pale grayish green, with a glittering white bloom on new growth and undersides of leaves. Plants have prostrate stems, strongly erect at the tips, that spread quickly into a loose, vigorous ground cover up to 1½ feet tall and 5 feet wide (45 cm by 1.5 m).
**Ornamental Drawbacks:** Flowers are inconspicuous and growth is somewhat rambling.
**Adaptability:** New Zealand spinach needs warm soil for germination, which takes several weeks. Soaking the hard seeds and sowing them early indoors help to produce an early crop. Plants grow fast in full sun or partial shade. Harvest begins about two months from seed and continues through the hottest summers. At each harvest, you can remove the top 3 to 4 inches (8 to 10 cm) of growth. Plants respond by branching and growing vigorously. They do fine in both cool and hot weather and survive light frosts, producing new leaves through fall in mild climates.
**Insects and Diseases:** There are no serious problems. Leaf miners and slugs occasionally do some damage.

3-71. New Zealand spinach. (From Mm. Vilmorin-Andrieux's *The Vegetable Garden.*)

**Food Uses:** Use New Zealand spinach leaves fresh or frozen, steamed or stir-fried. They have a pleasant, mild flavor. However, if New Zealand spinach leaves are eaten raw, the small hairs on the undersides of the leaves may irritate some people's throats.
**Landscape Uses:** New Zealand spinach is an attractive ground cover through summer and early fall. It quickly recovers its good looks after every harvest. Grow it in beds one plant wide and situate it so that you can reach into the plant mass from several sides to harvest. It hangs over the sides of raised beds and containers and softens their edges. If you like, encircle it with a colorful hedge of peppers and marigolds or with white-flowered, gray-leaved plants such as dusty miller. Nasturtiums and New Zealand spinach look excellent as adjacent ground covers and can be grown together on a bank or under light-canopied trees.
**Varieties:** None appears in seed catalogs.

## ***Squash, Summer (includes Zucchini, Yellow/Crookneck Squash, and Pattypan/Scallop) (*Cucurbita pepo*)
tender annual

**Ornamental Characteristics:** Summer squash are those whose immature fruits are used as vegetables. The leaves are bold-textured and up to 1½ feet (45 cm) long, lobed and cut to varying degrees, and toothed along the margins. (See figure 3-72.) They are medium or dark green, plain or mottled with silver patches. The deeply cut, silver-mottled leaves of zucchini varieties are the most attractive.

3-72. Zucchini summer squash. (Photo by Daphne Drury.)

Most summer squash grow in fountain-shaped bushes, narrow at the bottom and wide at the top, up to 5½ feet (1.7 m) across. Varieties differ greatly in plant size and erectness, and in leaf density. The finest-looking bushes are erect and densely leaved. The large orange flowers are often hidden in foliage, but are occasionally visible in blazing contrast to the leaves. Yellow and white fruits also contrast with the foliage.

**Ornamental Drawbacks:** Green zucchini fruits blend with leaves and have little ornamental value. Yellow and scallop squash plants have medium green, maplelike leaves that are not especially ornamental. In summer, leaves may become brown-edged and torn from weather and aging. Some varieties are not very erect. Others lose their bush form and trail along the ground. By late summer some varieties have so many stems that plants form tangled masses without much shape.

**Adaptability:** Summer squash does best in full sun, but is so vigorous and prolific that it produces well even in partial shade. Seeds germinate quickly in warm soil. Both male and female flowers are borne; some fruits are normally aborted. The first crop is ready about two months from seed, and from then on fruits come in quick succession. Prompt harvesting keeps plants productive.



**Insects and Diseases:** Downy or powdery mildew may be disfiguring in late summer, but these and other diseases do not usually injure squash plants seriously. Other disease possibilities are white mold, bacterial wilt, and squash mosaic virus. Insect problems include squash vine borers, squash bugs, cucumber beetles, squash beetles, melon aphids, white flies, melon worms, pickle worms, and slugs.

**Food Uses:** Use summer squash fresh, frozen, canned or pickled. Eat it raw in salads and hors d'oeuvres, or cook fruits whole when they are 2 to 6 inches (5 to 15 cm) long. Steam, boil, sauté, deep-fry, stir-fry, or bake them. Since the plants are prolific, it makes sense to eat the fruits while small and tender. Larger fruits need to be sliced or grated or stuffed and baked. You can fry male flowers and roast ripe seeds of mature fruits as you would small pumpkin seeds.

**Landscape Uses:** Grow squash plants alone, in clumps, or in rows. When you grow plants together in hills, they lack a distinctive shape; they will be more effective if spaced several feet apart. Just be sure to leave room to harvest fruits without getting scratched by the bristly stems. Squash in a big tub grows into a lush, tropical-looking patio plant.

Squash looks good with tall, bright flowers—tall marigolds, zinnias, cleome, amaranthus, and dwarf sunflowers—planted near it or in among its stems. The latter arrangement works well if you give flower seeds a head start of several weeks before you plant squash. Vertical flower stems of hollyhocks, blue salvia, and snapdragons contrast well with the broad growth habit of squash. A grape or cucumber trellis, or a wall of corn, sunflowers, or sorghum forms a strong background. Bright-flowered or bright green foliage plants are effective in the foreground. Examples are dwarf marigolds, geraniums, celery, and peppers, which are plants not so small or fine-textured that they are overwhelmed by the squash.

**Varieties:** Almost all zucchini varieties are attractive to varying degrees. They have little or no powdery mildew. Leaf color, leaf shape, and plant growth habit are best in 'Blackini,' 'Castle Black,' and 'Castle Verde,' closely followed by 'Diplomat,' 'Aristocrat,' and 'Burpee Hybrid.'

'Eldorado,' 'Gold Rush,' and 'Golden Eagle' are handsome yellow-fruited varieties. 'Eldorado' is a big plant with very large leaves, variegated and deeply lobed like those of zucchini. The flowers are clearly visible, and the delicious fruits are zucchini-shaped and glowing yellow. 'Gold Rush' resembles 'Eldorado.' 'Golden Eagle' remains erect and untangled all summer. Fruit is crook-necked and yellow; attractive leaves are maplelike with crisp edges, but may become covered with powdery mildew in late summer.

Scallop or pattypan squashes are no beauties. The most pleasant-looking is 'Scallopini,' a zucchini-scallop cross with dark green, scallop-edged fruits, and maple-lobed leaves with rounded tips. It has little powdery mildew, and an erect habit it loses in late summer.

'St. Pat. Scallop,' 'Creamy,' and 'Goldneck' are unattractive squash varieties.

3-73. Winter squash.

### ***Squash, Winter, and Pumpkin (varieties from *Cucurbita pepo, C. maxima, C. moschata,* and *C. mixta*)
tender annuals

**Ornamental Characteristics:** Winter squash and pumpkins are distinguished from each other by food use, rather than by species. Mature, hard-shelled fruits of winter squash are used as vegetables. The flesh is fine-textured and mild-tasting. Pumpkin flesh, on the other hand, is coarse-grained and strong-tasting. It is seasoned and cooked in pies, rather than used as a plain vegetable. Pumpkins are used as jack-o'-lanterns and livestock food as well.

Winter squash and pumpkins usually grow into big, vigorous vines that form a large mass of interwoven branches, or a dense central mass with long runners ambling outward. A few varieties grow in erect bushes.

Leaves are up to 15 inches long and 15 inches wide (38 by 38 cm) and margins are toothed. (See figure 3-73.) Leaf shape may be rounded and scarcely lobed at all or cut into five lobes like maple leaves. Leaves of some pumpkins are deeply lobed like those of zucchini squash. Color is gray-green, medium green, dark green, or green with silver mottling. Most leaves are matte-surfaced; a few are glossy.

Winter squash and pumpkins have large, bright orange flowers, more or less visible depending on varietal growth habit. Fruits have a wide range of sizes, shapes, surfaces, and colors—from 4 inches (10 cm) long to several feet, from 1 pound to 350 pounds (.5 to 159 kg), in spherical, oval, turban, pear, apple, or acorn shapes; smooth, grooved, or warty; green, yellow, orange, white, gray, blue-gray, or black; speckled, striped, solid, or bicolored. The flesh is usually yellow or orange and seeds may be thin or plump, naked or covered with a hard seed coat.

**Ornamental Drawbacks:** The plants take up a lot of space! However, there are several ways to control their space hunger. You can train wandering runners back into the central plant mass; trim off some of the runners that will produce fruit too late to ripen in your climate; or grow them vertically.

There are additional drawbacks as well. Some varieties are less dense than others and more ungainly—thin in the middle with long, stringy runners. Some have leaves of undistinguished color and shape. Many have disfiguring leaf diseases or become badly weathered by late summer. Flowers and fruits are usually well hidden in foliage.

**Adaptability:** Winter squash and pumpkins are slow to produce ripe fruits. They take 85 to 130 days depending on variety and climate. In cool springs or short summers, you may need to start them early indoors or to start them outdoors under paper hot caps. Seeds germinate quickly and seedlings grow fast in warm soil.

Although plants may survive drought, the fruit quality will be lowered. Fruits mature best in warm, dry weather. During rainy autumns, prop them off the wet soil to avoid rot. You can harvest fruits before or after frost, although the vines are killed by frost.

**Insects and Diseases:** These are the same as for summer squash, but with one difference. If summer squash plants become diseased and ugly in late summer, you have already harvested plenty of fruit and you can pull them up without feeling bad. Winter squash or pumpkin plants may become just as diseased, but must be kept until fruits ripen.

**Food Uses:** Bake, steam, purée, fry, or stir-fry winter squash. Make pumpkins into pies, cookies, puddings, soups, jams, chutneys, and wines. Actually, you can use winter squash for most of the same purposes as pumpkins, but its taste will not be so definite, nor its flesh so well suited to long cooking. Winter squash and pumpkins are excellent sources of vitamin A and contain moderate amounts of protein and iron. There are subtle variations in flavor and texture among different varieties. You can can them, freeze them, or store them whole at cool temperatures for two to four months after you have cured them for a week at about 80° F. (27° C.).

You might also fry the male flowers and roast or fry the seeds, with or without seed coats. While delicious-tasting and good sources of protein, minerals, and B vitamins, seeds are high in fat.

**Landscape Uses:** Plant squash and pumpkin vines near early-summer lettuce or early-blooming annuals. When you pull out the early plants, the vines fill any gap left in the garden. Grown vertically on fences, trellises, and summer houses, their bright flowers and varied fruits are more visible; extra-heavy fruits may require support from slings or platforms. Plants can grow lushly beside a pool. You can plant them in rows, hills, or vast rambling beds. They thrive in places that might not otherwise be used—on stumps, compost heaps, and tool sheds, in vacant lots, in border strips between the sidewalk and street. (Pumpkins, with their inviting orange fruit, should not be grown near the street, however.)

Large, bright flowers on stocky, big-leaved plants can be grown as a border for a big squash bed. Calliopsis is relatively delicate in texture, but so strong-colored that it holds its own visually. Tall sunflowers, corn, and sorghum are excellent background plants for squash and pumpkins; tithonia works well in back of gray-leaved squash. When squash or pumpkins are grown vertically, few flowering vines can keep up with them. In climates where they are well adapted, morning glories may do well if given a long head start.

**Varieties:** The best-looking squash in our trial gardens was 'Delicata,' also known as 'Sweet Potato Squash.' 'Golden Delicious,' 'Vegetable Gourd,' 'Royal Acorn,' and 'Triple Treat' were also very attractive.

'Delicata' is an old New England variety. Its medium green leaves have a good deal of fine silver mottling; leaf edges are crisply toothed, and sharply lobed like those of a maple. The vine is dense, up to 2½ feet tall and 16 feet wide (75 cm by 4.8 m). Its tendency to get powdery mildew is slight compared to that of many other varieties. Its oval fruits are cream-colored with green stripes. They store extremely well and taste mild, fine-grained, and slightly sweet.

'Golden Delicious' has soft, rounded leaves with a slight tendency to get powdery mildew. It forms a 10-foot (3-m) central plant mass with long side runners. Its large, lumpy orange-red fruits are easy to see. The coarse-textured, strong-tasting flesh is often used in pies.

'Vegetable Gourd' has gray-green, round-lobed leaves with a middle-range tendency to get powdery mildew. 'Vegetable Gourd' grows into a dense mass about 7 feet (2.1 m) wide, with short side runners. The small fruits are cream-colored with green spots and stripes, round with an indented top and deeply grooved sides. The flesh is fine-grained and pleasant.

'Royal Acorn' has maple-lobed leaves of dark, glossy green, with a slight tendency to develop powdery mildew. It is a broad, dense vine with slightly elongated acorn squash fruits.

'Triple Treat' pumpkin has dark green leaves of two different shapes—deep-lobed like zucchini, and maplelike with rounded lobe tips. They become weathered by early fall, but are only slightly disfigured by powdery mildew. 'Triple Treat' is relatively small and has a rather open habit. Male flowers are often carried above the foliage; the orange fruits are easy to see. They can be used as jack-o'-lanterns and taste excellent in pies. Their seeds are large and devoid of seed coats.



The least attractive winter squash and pumpkin varieties grown were butternuts—'Waltham Butternut,' 'Early Butternut,' and 'Bush

3-74. Bush winter squash 'Gold Nugget.'

Butternut' (not a true bush). Fruits are of good eating quality, but leaves and growth habits are undistinguished. Plants become badly weathered by late summer. 'Funny Face,' 'Lady Godiva,' and 'Vegetable Spaghetti' are also poor ornamentals, somewhat attractive in early summer, but deteriorating rapidly later on. 'Vegetable Spaghetti' has unusual, spaghettilike flesh and may be worth growing in an inconspicuous part of the garden.

There are three beautiful, rampant, vining cucurbits whose uses are uncertain—'Huicha,' 'Black Seeded Gourd,' and *C. ficifolia.* The first two take a very long time to produce and ripen fruit and may not ripen any in cool or short summers. Suggested uses for *C. ficifolia* (which translates appropriately as "fig-leaf squash") are sweet-pickling the flesh and roasting the seeds.

Four bush varieties of winter squash tested in our gardens looked beautiful in midsummer, but by early fall the two acorn squash—'Table King' and 'Bush Table Queen'—had badly weathered leaves, and 'Emerald,' a buttercup, had severe powdery mildew. 'Gold Nugget' was the only one still in good condition. (See figure 3-74.) It had rounded leaves with slight powdery mildew, a long-lasting bush habit, and a prolific crop of round, yellow-orange fruits. In late summer ripening fruits tasted delicious in stir-fries, but when fully ripe the flavor was bland.

3-75. Sunflower.

### ***Sunflower (*Helianthus annuus*)
half-hardy annual

**Ornamental Characteristics:** Sunflowers are a massive presence in the garden. Sometimes over 10 feet (3m) tall, they are heavy-stalked, large-leaved, and large-flowered. Rough-surfaced, gray-green leaves that are heart-shaped with prominently toothed edges measure up to 1½ feet long and 1½ feet wide (45 by 45 cm). (See figure 3-75.) Enormous composite flowers turn toward the brightest sun. Ray flowers are usually yellow, and central disc flowers brown. Disc flowers later grow into a geometric pattern of black and white seeds. Since sunflowers usually have only one flower per stem and flowers are round and face-sized, the plants have a definite peoplelike quality; it gives them a charming, humorous look.

**Ornamental Drawbacks:** Plants are rather coarse and do not look like much until they begin blooming. There is one main stalk, often dragged down by the weight of the flowers. Erect habit can be encouraged by wide spacing so plants will grow stocky, by tying the plants to a fence or other support, and by deep spading of the soil before planting to allow extensive rooting.



**Adaptability:** Sunflowers are native to the western and central United States. They do well in hot or cool summers and grow in a wide range of soils. Seeds germinate well from mid-spring on; seedlings grow quickly. Ripening seeds may need protection from birds. You can cut the flowerheads when the outermost seeds are ripe and dry them indoors.

**Insects and Diseases:** Insects include sunflower moth larvae, sunflower maggots, stalk borers, and sap-sucking plant bugs. Powdery mildew, sclerotinia, and rust are possible diseases. 'Sundak' is rust-resistant.

**Food Uses:** Eat sunflower seeds raw or roasted; add them to salads and baked goods; sprinkle them on desserts, soups, vegetables, and casseroles; or grind them into meal and use them in baking. John Gerard (See Bibliography) recommends sunflower buds in his herbal, "boiled and eaten with butter, vinegar, and pepper . . . an exceedingly pleasant meat, surpassing the artichoke far in procuring bodily lust." Whole seeds can be fed to poultry and wildlife or roasted and brewed as a coffee substitute.

**Landscape Uses:** Sunflowers are excellent temporary hedges, screens, and space dividers. Grow them with other tall plants, such as corn and sorghum. They can be silhouetted against a fence, hedge, or vine trellis, or planted in back of it so their flowers peer over the top.

Make sure plantings are arranged so the sunflower heads will face toward the center of the garden.

Big, stocky plants in front of sunflowers help mediate their scale to the rest of the garden; they hide the plainness of young sunflower plants as well, and cover up the messy stem bases. Summer and winter squash and pumpkins, big zinnias, black-eyed Susans, dwarf sunflowers, and amaranthus plants are attractive companions for sunflowers.

**Species and Varieties:** Some sunflowers have large seeds developed for people to eat. In our trial gardens, 'Mammoth' (= 'Mammoth Russian') plants grew tall and fairly uniform in height, 5 to 6½ feet (1.5 to 2 m) tall with 10- to 15-inch (25- to 38-cm) flowerheads. 'Sundak' plants were 3½ to 6 feet (1.1 to 1.8 m) tall with 12-inch (30-cm) flowerheads. 'Sundak' bloomed one week before 'Mammoth.'

Other varieties have been developed as oil sources and ornamentals. These generally have small seeds better suited for birds than for people. Small-seeded sunflowers usually are found in flower sections of seed catalogs. These plants often grow smaller, branch more, and flower more abundantly than large-seeded varieties. Flowers usually are smaller, single or double, red, brown, gold, yellow, or white.

Jerusalem artichoke (*Helianthus tuberosus*) is listed under "Less Attractive Vegetables and Herbs" at the end of this chapter. American Indians ate the tubers of *H. doronicoides,* which resembles Jerusalem artichoke, and they ground the seeds of perennial *H. giganteus* for flour.

3-76. Swiss chard. (From Mm. Vilmorin-Andrieux's *The Vegetable Garden.*)

## ***Swiss Chard (*Beta vulgaris,* Cicla Group)
hardy biennial often grown as an annual

**Ornamental Characteristics:** Swiss chard is a large-leaved, narrow-rooted beet. One Swiss chard variety, 'Rhubarb,' occasionally known as 'Ruby,' has glossy green leaves with leaf stalks and veins of shining deep red. In cool weather, the green part of the leaves changes to reddish brown. The colors of this plant are its star attraction.

Leaf margins are slightly wavy and surfaces are ornately blistered. Erect leaf stalks grow from the crown of each plant; older stalks at outer plant edges bend over as they age. (See figure 3-76.) Plants may become 3 feet tall and 2½ feet wide (90 by 75 cm); their effect is strongly vertical.

**Ornamental Drawbacks:** Other Swiss chard varieties lack the bright coloring of 'Rhubarb' and are not especially ornamental. They have shiny green leaves, sometimes blistered, with white or yellowish white leaf stalks and veins. Plants are sometimes wider than tall, and usually less erect than 'Rhubarb.'

'Rhubarb' chard sometimes produces tall, unattractive flower stalks in late summer. It is slightly less cold-hardy than other varieties.

**Adaptability:** Seeds can be sown from early spring on. Keeping soil moist may encourage good germination, but be prepared for surprises. Some seeds do not sprout at all and others sprout into plant clusters. Plant thickly; if germination is uneven, you can move seedlings around.

Once Swiss chard sprouts, it is easy to grow in hot or cool weather, full sun or partial shade. Plants are stouter in full sun. Harvest of outer leaf stalks begins about two months from seed. Repeated cuttings encourage new growth, and harvest continues through summer

3-77. French (winter) thyme.

and fall. Plants survive light frosts, and last a long time even in cold climates if grown in a heavy mulch or cold frame.

**Insects and Diseases:** Possible insects include aphids, flea beetles, beet leaf miners, snails, and slugs. Disfiguring leaf diseases include scab, mildews, and leaf spots. They are usually not serious, however.

**Food Uses:** Use Swiss chard leaves as you would spinach, raw or cooked, fresh or frozen. You can cook young stalks with the leaves, but older stalks are tougher and should be cooked separately.

**Landscape Uses:** 'Rhubarb' Swiss chard can be planted in rows, small groups, and masses, in beds, borders, and containers. You can combine it with red-leaf lettuce, beets, 'Dark Opal' basil, red-leaf amaranthus, and red-leaf corn. It contrasts well with the bright greens of parsley, celery, carrots, and some lettuce varieties. Its blistered leaves go well with either bold or fine-textured plants and with many different flower colors—deep red, yellow, orange, and white. In the fall when the foliage bronzes, it looks best with lavender and pink flowers.

**Varieties:** Green-leaved varieties include 'Lucullus,' 'Fordhook Giant,' 'White King,' 'Perpetual,' and 'Common Green.' The first three have blistered leaves and prominent white stalks and veins. The last two have plain leaves and less prominent veining. 'Burgundy Giant' is a red-stalked variety worth trial.

## ***Thyme (*Thymus* species)
hardy perennials

**Ornamental Characteristics:** Most of thyme's characteristics are variable. Leaves may be broadly ovate, or narrow and almost triangular, dark green to pale gray; thin and smooth; leathery and rather thick, or wooly. (See figure 3-77.) They are dense and borne in regular pairs along the stems, giving plants a fine, even texture. They are often evergreen, turning reddish in cold weather. Growth habits range from creeping mats to stiff, erect bushes. Delicate flowers bloom in short, rounded spikes of purple, lavender, red, pink, or white. Flowering begins from early summer to midsummer and sometimes continues for several months.

**Ornamental Drawbacks:** Thyme grown in shade or partial sun may become leggy. The lower stems on older plants of creeping thyme may become woody; they drop their leaves and then the tangled, bare stems are visible through new growth above.

**Adaptability:** Thyme grows best in full sun and well-drained soil, but tolerates shade and poor drainage. Many kinds can be grown from spring-sown seed. Cuttings and division are other propagation methods. Seeds germinate quickly, but seedlings grow slowly at first. A head start indoors produces bigger plants that may bloom in late summer the first year. Some thymes self-sow and spread themselves around the garden. Creeping thyme spreads rapidly along the ground, but can easily be controlled. Many references suggest renewing thyme plantings after three or four years.

**Insects and Diseases:** Thyme is very healthy.

**Food Uses:** Once established, you can harvest thyme at any time. Plants that have been cut back early will flower in late summer. Use thyme fresh, frozen, or dried, to season meat, poultry, stuffing, sausages, soups, stews, bean dishes, cheese, and pickles. Thyme is used in herbal medicine as a remedy for sore throats, stomach upsets, fevers, and headaches; bees make strong, pleasant honey from the flowers.

**Landscape Uses:** Neat bushes of upright thyme can be lined up as edging plants, massed in beds and borders, or planted in pots and window boxes. You can combine them with other plants of similar texture, such as rosemary, lavender, and summer and winter savory, or of related colors, such as kale, lima bean, rue, and dusty miller. Dianthus, lobelia, pimpernel, linaria, and alyssum have fine-textured foliage and flowers that agree with thyme's lavender blooms. In spring, the foliage is a good background for grape hyacinth and dwarf daffodil.

Creeping thyme makes a strong-growing ground cover along paths and lawn edges, up hills and banks, down retaining walls, and over the edges of pots, window boxes, and hanging baskets. You can plant it at the base of bushes, in rock gardens, in beds, in flower borders, and in planting strips between sidewalk and street. You might let it grow between flagstones in a path, although its growth will be slow if it gets stepped on often.

You can grow one creeping thyme variety alone, or combine different types if they are equally vigorous. Tall creeping thyme varieties and oregano thrive together in full sun.



**Species and Varieties:** English and French thyme (both *Thymus* *vulgaris*) are erect bushes up to 1 foot tall and 2 feet wide (30 by 60 cm). Both are widely used in cooking and herbal remedies. English thyme has relatively broad leaves that are red underneath; it is not commonly grown from seed. French (or winter) thyme (shown in figure 3-77) has prominent, wiry stems and narrow, gray-green leaves curled under at the edges. It looks like a desert plant, tastes stronger than English thyme, and can be propagated from seed, cuttings, or division. Both English and French thyme have pink flowers in June.

Seeds labeled 'Winter Thyme' and 'French Summer Thyme' produce plants that look the same. 'Winter Thyme Dwarf Compact,' smaller and more densely leaved, was the most attractive *T. vulgaris* variety in our trial gardens.

Creeping thyme or mother-of-thyme (*T. serpyllum*) grows fast as a ground cover up to 4 inches (10 cm) tall, forming even-textured mats several feet across. Small, ovate leaves are clear, dark green; stems are reddish brown. Plants bloom in August the first year from seed and continue for six weeks or more. In successive years they may bloom off and on all summer. The rose-lavender flowers, in small clusters on upright stems, seem to float above the leaves. Creeping thyme tastes milder than common thyme, but can be used in the same ways in cooking and herbal remedies. Its many varieties include white-flowered 'Albus,' red-flowered 'Coccineus' and 'Splendens,' pink-flowered 'Roseus' and 'Pink Chintz,' silver thyme with white-variegated leaves, and 'Aureus' with yellow-variegated leaves.

Other creeping thymes include lemon thyme and wooly mother-of-thyme. Lemon thyme resembles ordinary creeping thyme in all but its scent; it tastes good in salads and with fish. Wooly mother-of-thyme, with tiny, gray, hairy leaves and lavender flowers, requires very good drainage.

*T. caespititius* has a tangerine fragrance and forms mosslike hummocks. Caraway thyme (*T. herba-barona*), a dense ground cover with narrow, dark green leaves and purple midsummer flowers, is an excellent cooking herb.

## *Tomato (*Lycopersicon lycopersicum=L. esculentum*)
tender perennial grown as an annual

**Ornamental Characteristics:** Tomatoes have advantages as ornamental food plants, but also distinct drawbacks. Choice of variety can be very important. Plants may be compact, erect bushes 1 foot tall and 2 feet wide (30 by 60 cm). But they may grow in low, straggly mats, or in big masses up to 5 feet tall and 15 feet wide (1.5 by 4.5 m). They may be determinate—self-topping, with fruit that ripens in a short period of time—or indeterminate, continuing to produce new vegeta-

**3-78.** Tomato.

tion and fruit until killed by frost. Leaflets may be dark green or medium green, bold-textured and deeply veined, medium-textured, or fernlike and delicate.

The yellow flowers are attractive, but so small in relation to plant size that they are effective only on small plants. Fruits are red, pink, orange, yellow, or white, and round, broad oval, pear-shaped, or plum-shaped. Their surfaces are smooth or grooved. They measure from 7/8 inch to 4½ inches (2 to 11 cm) in diameter. They look like bright Christmas balls when they ripen among healthy green leaves.

The most attractive tomato varieties have bright-colored fruits, and fine or bold-textured foliage that stays green and healthy-looking all summer.

**Ornamental Drawbacks:** Rampant-growing tomatoes are hard to use in small gardens. Large varieties can be caged, trained on trellises, or turned loose to romp over large spaces. Many tomato leaves are medium-textured and not especially ornamental.

Leaf curl is minimal in some environments, but it can be a big ornamental problem. Leaves of many tomato varieties curl up and look as if they are wilted or dying. This happens as fruit turns color, just when the plants' ornamental qualities should theoretically be at their peak. Shiny red fruits are borne on diseased-looking plants. Leaf curl is not a disease, however; it is part of the normal physiological processes of tomatoes. Heavily pruned plants or those with a small leaf-fruit ratio have more leaf curl. A few varieties have a genetically determined fondness for curling their leaves and begin doing it long before fruit ripens.

**Adaptability:** In most parts of the country tomatoes should be started indoors about six weeks before the weather is warm. Where summers are very long, they can be seeded outdoors in late spring. Pinching off sideshoots gives earlier yield, but unpruned plants provide leaf cover to prevent sunburn and have greater yield in long summers. Flowers set fruit best when night temperatures are above 55° F. (13° C.) and below 75° F. (24° C.). Flowers may drop if temperatures are too high or too low, but more will bloom later on. Late-setting fruit may not ripen in short summers. Most tomatoes develop the best-flavored fruit in warm, dry weather. Training plants vertically or using clean mulch under prostrate plants will reduce fruit rot and cracking in wet weather. Plants often reseed themselves.

**Insects and Diseases:** Possible insects are white flies, red spider mites, flea beetles, blister beetles, Colorado potato beetles, aphids, cutworms, fruitworms, hornworms, pinworms, leaf miners, and leaf hoppers. Some insects cause direct damage to tomato plants and others are important because they carry disease. Diseases include fusarium wilt, verticilium wilt, nematodes, early blight, late blight, anthracnose, mosaic, and other diseases important locally. Catalogs often list variety resistance to the three most widespread diseases—verticilium wilt, fusarium wilt, and nematodes—after variety name: for example, 'Small Fry' VF.

Your tomato plants will not have every insect and disease on this list, but two or three different problems may come up during the growing season. In addition, physiological problems of tomato fruits are blossom end rot and fruit cracking.

**Food Uses:** The many uses of tomatoes are well known. Tomato flavors range from bland to strong and sweet. Textures range from meaty (and sometimes mealy) to juicy and possibly seedy, while fruit skins may be soft or tough. Large, meaty tomatoes are most often used for slicing in salads and sandwiches. Cherry tomatoes, used whole in salads and sauces, are juicier and stronger-tasting. Their flavor is less affected by poor weather, but may turn musky if fruits become overripe. Their skins soften during cooking.

Indeterminate tomatoes that produce over a period of time are best for fresh use; use determinate varieties if you plan on canning and need a lot of fruits that will ripen at about the same time. Most varieties are acid enough (below pH 4.6) to avoid danger of botulism when they are canned. Picking tomatoes before they are overripe and adding one tablespoon of lemon juice per pint (.47 liter) of canned tomatoes are good safeguards.

**Landscape Uses:** Plant small bush tomatoes singly or in groups and rows. They look attractive, too, in containers, in beds, and in borders. Medium-sized, small-fruited vining varieties can cascade from hanging baskets. Big vines growing loose make vigorous covers for sunny banks and compost heaps. They can also grow in wire cages or on trellises and openwork fences.

Tomatoes must be tied to their supports, but plants do not look good tied to a single stake with all their side shoots pinched out. Keep in mind that tomatoes grown near the street invite snacking and throwing. You may not mind sharing, but inexperienced pickers are likely to accidentally pull up a plant by its roots. Do not grow tomatoes where very small children will be playing unsupervised, because even small amounts of tomato leaves and stems are quite poisonous.

Cherry tomatoes make good "potluck" plants for community gardens. You can offer a few to neighborhood people who drop by the garden just to see what's going on. This friendly sharing discourages vandalism and helps make neighborhood people protective of your community garden.

Fine-leaved tomatoes look best with small-leaved, bright-flowered plants such as salpiglossis, and cosmos 'Diablo' with its red-orange flowers. Nearby fine-textured parsley or carrots are also effective. Grow bold-leaved tomatoes with stockier companions such as zinnias, marigolds, and geraniums. Surround tomato cages with flowers. Trellises can be shared with vigorous flowering vines such as morning glory.

**Species and Varieties:** We grew about sixty tomato varieties in our trial gardens. None looked great under the environmental conditions we had and all had leaf curl to some extent. The plants were grown on the ground.

Some cherry tomatoes—'Red Cherry,' 'Red Cherry Small,' 'Small Fry' VF, 'Yellow Plum,' and 'Gardener's Delight'—were fairly attractive. Fruits are usually borne in clusters high on the plant, where they are clearly visible and less likely to rot. 'Small Fry' ripens earlier than the others. Plants are dense, indeterminate, and usually quite big.

Paste tomatoes—'San Marzano,' 'Royal Chico' VF, and 'Roma' VF—also did well. Their oval fruits have meaty flesh well suited to canning and sauces. Plants are determinate and growth is fairly compact and erect. Leaflets are large and dark green.

Many varieties tested had fruits of standard tomato shape and size. Only four did well as ornamentals—'Heinz 1370,' 'Floramerica' VF, 'Redpak' VF, and 'Bonus' VFN. Plants are fairly compact, not especially erect, and fruits often rest on the soil. 'Bonus' and 'Redpak' have attractive leaflets that are large and dark. 'Heinz 1370' and 'Redpak' bear fruits up to 4 inches (10 cm) in diameter. None of these varieties ripen early.

3-79. Sweet violet.

Erect 'Dwarf Champion,' sharp-leaved 'Kootenai,' and 'Saladmaster,' with blue-green foliage, are attractive tomatoes that do not fit easily into the categories above. Some dwarf and very early varieties tested had severe leaf curl and discoloration. 'Tiny Tim,' 'Pixie,' 'Presto,' 'Early Subarctic,' 'Jetfire,' 'Springset,' and 'New Yorker' were among the worst-looking.

A wild tomato, *L. pimpinellifolium,* has delicate-textured foliage and very small, edible fruits.

## **Violet/Sweet Violet (*Viola odorata*)
hardy perennial

**Ornamental Characteristics:** Sweet violets grow in low mounds about 6 inches (15 cm) tall. They spread by seeds and runners to form pleasant ground covers. Leaves are glossy medium green, heart-shaped, and scallop-edged. The fragrant spring flowers have five purple petals, white at the base, with tiny yellow flower parts at each flower's center. Some varieties have lavender, pink, or white flowers, and some are double rather than single.
**Ornamental Drawbacks:** Violets are not showy when they are not blooming.
**Adaptability:** Violets grow well in ordinary soil. If the soil is rich, they are very leafy and produce few flowers. If it is too wet, they may become diseased. They need partial shade from hot summer sun.

They can be propagated from seeds or from rooted runners they send out after flowering. Vigorous single-flowered violets often produce as many blooms on new plants as they did on the original ones, but double-flowered violets may not, and runners should be removed. If plants become too crowded, flowering slows down. Plants may die back to the ground in fall and send up new leaves early in spring. In mild climates or where protected by mulch, they are evergreen.

**Insects and Diseases:** Possible insect pests include red spider mites, violet sawflies, violet gall midges, cutworms, and slugs. Anthracnose, scab, powdery mildew, root rot, crown rot, and wilt are possible diseases. If you buy violets at a nursery, check to see what they have been sprayed with so you know whether you will be able to eat them.

**Food Uses:** Violet flowers are a good source of vitamin C and you can eat them raw in salads and garnishes. They also add flavor and purple coloring to vinegar. They can be crystallized in sugar or made into violet jam, jelly, wine, or syrup. Violet flowers and syrup have been used in herbal medicine to relieve constipation and other ills and to make poultices and ointments for skin ailments.

Violet leaves are excellent sources of vitamins A and C. When they are very young, add them to salads, or steam them as potherbs with other greens. They are slightly acidic and mucilaginous. Violet leaves become more gummy as they age and can be used to add body to soups and stews. A native American violet, *V. palmata,* is commonly known as "wild okra" in the southern United States, where it is used to thicken stews.

**Landscape Uses:** Violets are not good specimen plants. They have a mild, quiet beauty that does not lend itself to being spotlighted. They are best as ground cover plants—the more the better if you plan to make violet syrup or jam. They do well under light-canopied deciduous trees and shrubs. They are an early sign of spring, purple and vivid green among last year's fallen brown leaves. You can plant them alone or with spring-flowering bulbs.

**Species and Varieties:** Single-flowered sweet violets are often more vigorous than double ones such as 'Parma.' Single-flowered varieties include 'Princess of Wales' and 'Royal Robe.'

There are many other violets besides sweet violets, and not all are used as food. Books of wild food plants suggest wild American violets of various species as sources of edible flowers and leaves. Euell Gibbons (See Bibliography) singles out *V. sororia* (=*V. papilionacea*) of the northeastern states. Some references claim that violas and pansies have edible flowers. Other references suggest that they have harmful properties, though I have not found reference to these in any books about poisonous plants.

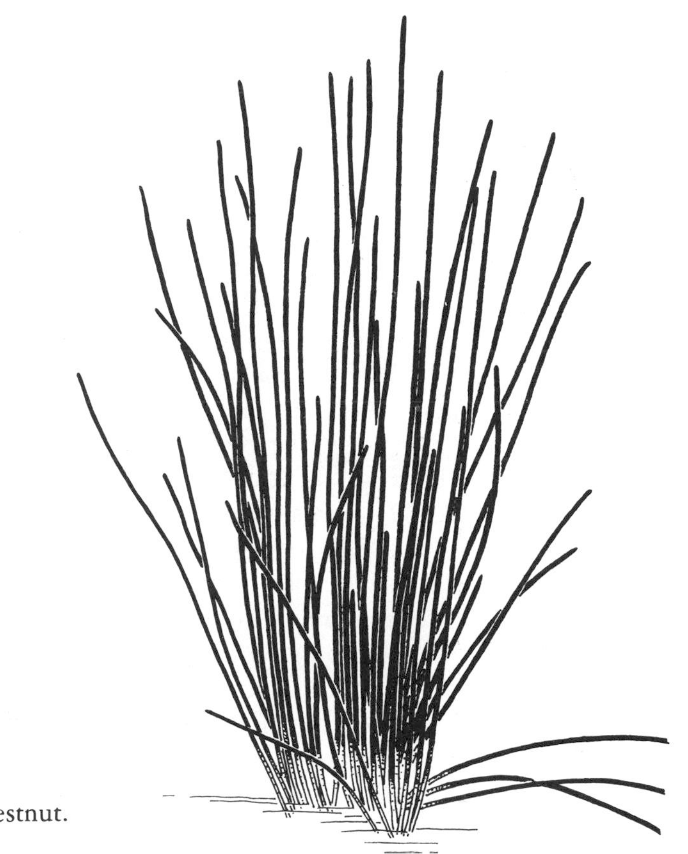

3-80. Water chestnut.

## **Water Chestnut (*Eleocharis dulcis*)
hardy perennial aquatic

**Ornamental Characteristics:** Water chestnut has long, needlelike leaves that rise out of the water in vertical clusters. (See figure 3-80.) They are pale, bright green and up to 2 feet (60 cm) tall.
**Ornamental Drawbacks:** The leaves are brittle. They break easily when people brush against them or children grab them, although they are seldom damaged by wind.
**Adaptability:** In the wild, water chestnuts grow at the edges of lakes and marshes. In China they are cultivated in rice fields or in large tanks. They are propagated from dark brown corms, available fresh from Oriental groceries or water plant nurseries. The corms are planted in early spring, underwater in large pots of very fertile soil or in the mud of a pond bottom. They grow best with only 1 inch to 2 inches (2.5 to 5 cm) of water above their soil, although they tolerate up to 6 inches (15 cm). They need full sun and do best in long, warm summers. In partial shade, their leaves topple over; they may grow slowly in cool weather. Each corm develops new corms on rhizomes around it; the new ones are ½ inch to 2 inches (1.3 to 5 cm) in diameter. They can be harvested or left to help the plants spread into a dense colony.

The plants will be more prolific and yield larger corms if they have plenty of room to grow.

Corms survive winter as long as the soil they are growing in does not freeze. If it may freeze, they should be brought inside, potted in moist soil, and kept in a cool, dark place until spring.

**Insects and Diseases:** Water chestnuts are very healthy.

**Food Uses:** Peel, slice, or dice water chestnut corms and cook them in stir-fries, soups, and egg roll stuffings. Or, if you wish, eat them raw in salads. They have a crunchy texture and mild, nutty flavor. They are sold canned in grocery stores but taste better fresh. You can harvest them as needed during fall and winter.

**Landscape Uses:** The spiky leaves are a strong, effective contrast to round lotus and water lily leaves. They combine well with arrowhead and flowering rush and with their broad-flowered sedge relatives, papyrus and umbrella plant. Bunching onions and ornamental grasses look good nearby on dry land. Large-leaved plants, such as squash, are excellent contrasts. You can grow water chestnuts in water gardens of any size—barrels, tubs, ponds, the edges of lakes and slow streams.

**Species:** Caltrops (*Trapa natans*), listed under "Possibilities to Explore" at the end of this chapter, is sometimes called water chestnut.

## LESS ATTRACTIVE VEGETABLES AND HERBS

A number of vegetables and herbs have, in my opinion, few distinctive qualities to justify their use as ornamentals. There are others whose attractive qualities are combined with major drawbacks. Some of the plants listed in this section are coarse, weedy-looking, or just very ordinary. Some are attractive, but have very short life cycles. Others may be attractive when they are growing in the right environmental conditions, but had poor growth or disfiguring diseases in our Oregon trial gardens.

The plants listed here are best grown in an inconspicuous part of the garden. If space is tight, you may choose not to grow them at all, unless, of course, you like the food they produce very much or you like the way they look better than I do. This is not at all unlikely. You may find that some of your favorite plants are in this section; many of my judgments here are personal and the line that divides the plants in this section from those included in the Encyclopedia is in some cases (Jerusalem artichoke, cardoon, chervil, potato, for example) very thin.

**Anise (*Pimpinella anisum*)**
hardy annual

Wispy-looking; short-lived; seeds used as flavoring.

**Artichoke, Jerusalem (*Helianthus tuberosus*)**
hardy perennial

Attractive flowers; coarse, fast-spreading plant that can be used for a tall background plant; prone to mildew; produces large numbers of edible tubers that can be used in stews and stir-fries.

**Broccoli (*Brassica oleracea,* Italica Group)**
hardy annual

Big, plain leaves; many insects and diseases; flower buds cooked or eaten raw.

**Brussels Sprouts (*Brassica oleracea,* Gemmifera Group)**
hardy biennial

Plain leaves; plant structure looks like a broken umbrella; same insects and diseases that apply to cabbage family; small, edible heads or "sprouts" in leaf axils.

**Burdock/Gobo (*Arctium lappa*)**
hardy biennial

Enormous rosettes of heart-shaped leaves; small purple flowers that quickly become sticky burrs; the roots are peeled, boiled, and used in Japanese soups and stir-fries.

**Cardoon (*Cynara cardunculus*)**
hardy perennial

Big rosettes of jagged leaves; purple, thistlelike flowers; attractive varieties with deeply lobed leaves are less satisfactory as food plants; broad leaf stalks are blanched for several weeks in the fall, then peeled and cooked.



**Cauliflower (*Brassica oleracea,* Botrytis Group)**
hardy biennial often grown as annual

Big, plain leaves; many insects, diseases, and physiological disorders; "heads" of edible flower primordia can be eaten raw or cooked.

**Celtuce (*Lactuca sativa*)**
hardy annual

Distinct form of lettuce with open rosettes of oval, chartreuse leaves; leaves are often bitter; thick, succulent flower stems can be steamed, braised, or stir-fried.

**Chervil (*Anthriscus cerefolium*)**
hardy biennial

Rosettes of attractive, bright green, and deeply cut leaves; short life in most climates; flowers or dies back in warm weather; leaves are used as French seasoning.

**Chickpea/Garbanzo Bean (*Cicer arietinum*)**
tender annual

Finely divided, blue-green leaves; sparse, awkwardly branching bushes; dried seeds are cooked in soups, casseroles, and Middle Eastern foods.

**Chicory (*Cichorium intybus*)**
hardy biennial

Glossy, blistered leaves, plain oval or jaggedly toothed like those of dandelion; short-lived blue flowers; roots of some varieties used for coffee substitute and leaves of other varieties, including 'Witloof,' used in salads.

**Collards (*Brassica oleracea,* Acephala Group)**
hardy biennial

Big, oval leaves; same insects and diseases that apply to cabbage family; leaves cooked as greens and added to soups and stews; cold-tolerant winter vegetable.

**Corn Salad/Maché (*Valerianella locusta*)**
hardy annual

Oblong leaves that lie limply on top of each other in low rosettes; leaves for cold-weather salads.

**Cress, Garden (*Lepidium sativum*)**
hardy annual

Finely divided, bright green leaves; lacy, white flower clusters; short life; peppery young leaves added to salads.

**Cress, Upland/American (*Barbarea verna = B. praecox*)**
hardy biennial

Nondescript leaf rosettes and small yellow flowers; young leaves are good in salads.

**Dandelion (*Taraxacum officinale*)**
hardy perennial

Jagged-toothed leaves; short-lived, yellow flowers; roasted roots used as coffee substitute; crowns, flowers, and young leaves go in salads; flowers are used in wine; young leaves used as potherbs.

**Fava/Broad Bean (*Vicia faba*)**
hardy annual



Big plant with spotted leaves, white flowers, and branches that are often awkward-looking (See figure 3-81); seeds are cooked green-shell or dry; some people are allergic to the seeds.

**3-81.** Fava bean. (From Mm. Vilmorin-Andrieux's *The Vegetable Garden.*)

**Garlic (*Allium sativum*)**
hardy perennial

Sparse, thin leaves that are seldom erect; clusters of bulbs—"cloves"—are commonly used for seasoning.

**Horseradish (*Armoracia rusticana*)**
hardy perennial

Erect clusters of oblong, docklike leaves; attractive white flowers; fast-spreading plant that is hard to eradicate; roots are used in sauce and seasoning.

**Kailaan/Chinese Kale (*Brassica oleracea,* Alboglabra Group)**
hardy annual

Plain leaves and pleasant-looking white flowers; unopened flower buds, leaves, and stems are used in Oriental cooking.

**Kohlrabi (*Brassica oleracea,* Gongylodes Group)**
hardy biennial grown as annual

Odd-looking plant with scanty, blue-green leaves and turniplike enlargement on stem; the stem enlargement can be peeled and sliced raw for salads, cooked alone, or added to soups and stews.

**Kudzu (*Pueraria lobata*)**
hardy perennial

Sparse mats of long, wandering stems and beanlike leaves; the roots yield an edible starch.

**Lambsquarters (*Chenopodium album*)**
tender annual

Sparse, upright branches; triangular leaves; weedy flowers and seedheads; self-sows prolifically; young leaves are used in salads and potherbs; seeds can be ground into flour; its bright-colored relative, *C. amaranticolor,* is worth trial as an ornamental.

**Lentil (*Lens culinaris*)**
hardy annual

Feathery leaves; sparse bushes that died without producing seed in our Oregon trial gardens; dry seeds can be cooked in soups and casseroles—important in India.

**Marjoram/Sweet Marjoram (*Origanum majorana*)**
perennial that tolerates little frost and is often grown as an annual

Small, oval leaves; white flowers mostly hidden by green bracts surrounding them; wandering branches; leaves are used as seasoning.

**Melons (*Cucumis melo*)**
tender annuals

In our trial gardens we grew muskmelon or "cantaloupe," honeydew or winter melon, and mango melon.

Sparse vines with gray-green leaves, slightly lobed and toothed, and small yellow flowers; leaves turn yellow as fruit develops and plants look almost dead by the time it ripens; the fruit is a common treat.

**Mung Bean (*Vigna radiata = Phaseolus aureus*)**
tender annual

Fuzzy, beanlike leaves; sparse bushes with severe disease problems in our trial gardens; young pods can be stir-fried; dry seeds can be sprouted, cooked, or made into Chinese cellophane noodles.

**Onion, Common/Bulbing (*Allium cepa,* Cepa Group)**
hardy biennial grown as annual

Spiky, gray-green leaves that at first grow in erect clusters, then later fall over and look messy; bulbs are eaten raw or cooked.

**Potato (*Solanum tuberosum*)**
hardy annual, perennial by means of its tubers

Deep green, boldly cut leaves that are attractive when healthy but are often diseased; white, yellow-centered flowers; the tubers are cooked.

**Purslane (*Portulaca oleracea*)**
tender annual

Small, succulent leaves; stems erect or rooting along the ground; plants seed prolifically and become weeds; the leaves are cooked or eaten raw in salads.

**Rutabaga/Swede (*Brassica napus,* Napobrassica Group)**
hardy biennial grown as annual

Loose clusters of lobed, oblong leaves; subject to cabbage family insects and diseases; the roots are eaten raw or cooked.

**Salsify/Oyster Plant (*Tragopogon porrifolius*)**
hardy biennial grown as annual

Grasslike clusters of narrow, dull green leaves; purple flowers and elaborate, fluffy seedheads appear the second spring; the long roots can be peeled and then boiled, steamed, or added to soups and stews.



**Scorzonera/Black Salsify (*Scorzonera hispanica*)**
hardy perennial grown as annual

Clusters of plain, deeply veined leaves; yellow flowers that look like big dandelions; the roots are soaked and then boiled or fried.

3-82. Sorrel.

**Sorrel (*Rumex acetosa*)**
hardy perennial

Erect clusters of large oval leaves; tall stalks of docklike reddish flowers (see figure 3-82); tough plant, persistent in the garden; the sour leaves are used in soups, added sparingly to salads and potherbs; sorrel contains oxalic acid, to which some people are allergic.

**Tarragon (*Artemisia dracunculus*)**
hardy perennial

Russian tarragon is rangy, narrow-leaved, vigorous, and has little flavor; French tarragon is similar; it has less vigor, but its flavor is excellent and it is used as French seasoning.

**Turnip (*Brassica rapa,* Rapifera Group)**
hardy biennial grown as annual, or true annual

Coarse, hairy, gray-green leaves; subject to cabbage family insects and diseases; the roots are eaten raw or cooked; the leaves are cooked as potherbs.



**Watermelon (*Citrullus lanatus* = *C. vulgaris*)**
tender annual

Deeply lobed leaves and yellow flowers that are attractive through midsummer; plants turn yellow and dead-looking as fruit begins to ripen; the fruit is edible.

3-83. Spinach.

The vegetables and herbs that follow are, in my opinion, hopelessly ugly.

**Cumin (*Cuminum cyminum*)**
tender annual

Sparse, erect or sprawling bushes of threadlike leaflets; short life; the seeds are used as Mexican and Indian seasoning.

**Fenugreek (*Trigonella foenum-graecum*)**
tender annual

Distorted look; sparse, sprawling bushes; inconspicuous flowers and twisted pods with needlelike tips; short life; the seeds are used for seasoning, sprouts, or as a coffee substitute.

**Radish, Spring (*Raphanus sativus*)**
hardy annual

Loose clusters of prickly, gray-green leaves; subject to cabbage family insects and diseases; short life; the roots are eaten raw or cooked.

**Rocket/Roquette (*Eruca vesicaria* subspecies *sativa*)**
hardy annual

Upright rosettes of coarse leaves; messy-looking flowers and seeds; short life; the young leaves are eaten in salads and occasionally used as potherbs.

**Spinach (*Spinacia oleracea*)**
hardy annual

Glossy, dark green leaves, smooth or crumpled; low rosettes that are sparse with long-stalked leaves, or crowded-looking, as if the leaves have been squashed together on the plant; the leaves are eaten raw or cooked.

3-84. German chamomile with sage.

## PLANTS WITH LIMITED FOOD USES

This list includes plants grown for specialized purposes (such as teas), as well as plants whose flavor may appeal to only a few people, and plants that require extensive preparation before they can be used. Many of them make good ornamentals, but none is a great food producer. Most survive winters in many parts of the United States. Each is listed with its edible part.

**Akebia (*Akebia trifoliata* and *A. quinata*)**
Fruit has sweet, bland flavor.

**Bearberry (*Arctostaphylos uva-ursi*)**
Mealy fruit may taste better when cooked.

**Catnip (*Nepeta cataria*)**
Loved by cats; leaves and young shoots are used in tea.

**Chamomile, German (*Matricaria recutita*) (See figure 3-84) and Chamomile, Roman (*Chamaemelum nobile*)**
Flowers are used in tea.

**Darwin Barberry (*Berberis darwinii*)**
Fruit can be used in jam.

**Hackberry (several *Celtis* species)**
Small date-flavored fruits have large pits.

**Horehound (*Marrubium vulgare*)**
Leaves are used in tea and cough drops.

**Hyssop (*Hyssopus officinalis*)**
Leaves are used in tea and seasoning.

3-85. Sweet woodruff, with kale and lettuce. (Brooklyn Botanic Garden, Brooklyn, New York.)

**Maples (*Acer saccharum,* also *A. saccharinum, A. rubrum,* and *A. macrophyllum*)**
Sap is boiled down into syrup.

**May Apple (*Podophyllum peltatum*)**
Sweet, slightly musky fruit pulp can be made into jelly; the leaves, seeds, and roots are poisonous.

**Monkey-puzzle Tree (*Araucaria araucana*)**
Seeds are edible.

**Mountain Ash (*Sorbus aucuparia*)**
Small sour fruits can be used in jellies.

**Oaks (several *Quercus* species)**
Seeds of native American white oaks are edible after being boiled in several changes of water and then roasted.

**Pyracantha/Firethorn (*Pyracantha* species, especially *P. angustifolia* and *P. coccinea*)**
Mealy fruit is occasionally made into jelly.

**Russian Olive (*Elaeagnus angustifolia*); also *E. commutata* and *E. umbellata***
Fruit can be used in jellies.

**Sea Buckthorn (*Hippophae rhamnoides*)**
Sour fruits can be used in jellies.

**Strawberry Bush (*Arbutus unedo*)**
Insipid fruit can be made into preserves.



**Sumac (*Rhus copallina, R. glabra, R. trilobata,* and *R. typhina*)**
Hairy red fruit is steeped in water for a lemonadelike drink. (Do not confuse with white-fruited poison sumac.)

**Sweet Woodruff (*Galium odoratum* = *Asperula odorata*) (See figure 3-85)**
Leaves are used in tea; shoots are steeped in wine.

**Trifoliate Orange (*Poncirus trifoliata*)**
Bitter fruit juice is used as seasoning.

**Twinberry/Partridge Berry (*Mitchella repens*)**
Edible fruit is variously described as "spicy" and "insipid."

## POSSIBILITIES TO EXPLORE

There are a number of vegetables and herbs that I still need to learn more about. Although at this time I am still learning about their food uses and/or the conditions in which they grow best, I want to suggest them to you now because they are well worth exploring on your own.

All these plants have good possibilities as ornamentals. Some of them may be familiar to you from perennial borders and wildflower gardens; others are quite rare and difficult to obtain. Some are grown in many countries; others are cultivated in only a few. Some were once eaten by American Indians or Europeans and have recently fallen into obscurity. A few, such as sweet potato and peanut, are common food plants that do not grow well in the relatively cool summers of western Oregon where our trial gardens were located.

Most of them are winter-hardy in many parts of the United States. A few can be grown only during the summer months. Some are more adaptable than other food plants in terms of their light requirements and can be grown in partial shade. Many of the plants have a number of common names; they are listed here by their most frequently used common names. Plants that I consider especially attractive are marked by a star (*). Each plant is listed with its food use.

The information in this section came mostly from a few excellent sources: E. L. Sturtevant's *Sturtevant's Edible Plants of the World,* J. C. Th. Uphof's *Dictionary of Economic Plants,* Donald Wyman's *Wyman's Gardening Encyclopedia,* Julia F. Morton's *Herbs and Spices,* Mrs. M. Grieve's *A Modern Herbal,* and *Hortus Third,* which was initially compiled by Liberty Hyde Bailey and Ethel Zoë Bailey and revised and expanded by the staff of the L. H. Bailey Hortorium of Cornell University. (See Bibliography.) The last four of these sources are good starting points to gather information about appearance, culture, and specific food uses of the plants listed here.

**Adzuki Bean (*Vigna angularis* = *Phaseolus angularis*)**
Sweetish seeds are cooked green-shell or dried in casseroles and Oriental desserts.

***Angelica (*Angelica archangelica*) (See figure 3-86.)**
Stems and blanched leaves are cooked as vegetables; the seeds are used as seasoning.

**Asafetida (*Ferula foetida*)**
Young shoots and leaves are cooked as potherbs; gum that exudes when tops of roots are cut is used as Indian seasoning.

**Asparagus Pea/Goa Bean (*Psophocarpus tetragonolobus*)**
Pods can be steamed; dried seeds are roasted.

**3-86.** Angelica. (From Mm. Vilmorin-Andrieux's *The Vegetable Garden.*)

**Bamboos, some species: **Phyllostachys aurea, *P. aureosulcata, *P. bambusoides, *P. dulcis, *P. meyeri, *P. nigra, *P. nuda, *P. viridiglaucescens, *P. viridis, *P. vivax, Semiarundinaria fastuosa***
Boiled spring shoots, with bitter cooking water discarded, are added to soups and stir-fries, or preserved for later use.

***Bay Laurel (*Laurus nobilis*)**
Leaves are used as seasoning.

***Beebalm/Bergamot (*Monarda didyma*), *Lemon Bergamot (*M. citriodora*), and *M. fistulosa***
Leaves are used for tea and seasoning.

**Bellflower/Rampion (*Campanula rapunculus*)**
Roots, leaves, and blanched shoots are cooked as vegetables.

**Caltrops/Water Chestnut (*Trapa natans*)**
Seeds have been roasted, boiled, and eaten raw since Neolithic times.

**Canaf/Indian Hemp (*Hibiscus cannabinus*)**
Leaves are used as potherbs, and seeds as seasoning.

**Caper (*Capparis spinosa*)**
Spicy flower buds and young fruits are pickled and used in salads and casseroles.



**Caraway (*Carum carvi*)**
Seeds flavor cheeses and baked goods; roots can be boiled.

**Chayote (*Sechium edule*)**
In the Caribbean, the fruits are used like potatoes; the leaves and young shoots are cooked as potherbs; and the roots are fried or boiled.

3-87. Flowering rush.

**Comfrey (*Symphytum officinale*)**
Young leaves and blanched shoots are cooked as potherbs; plants have many herbal uses as well.

**Cowpea/Blackeye Pea (*Vigna unguiculata* subspecies *unguiculata*)**
Seeds are cooked green-shell or dry.

**Cuckoo Flower/Lady's Smock (*Cardamine pratensis*)**
Leaves are a substitute for watercress.

***Dogtooth Violet/Trout Lily (*Erythronium americanum*), *European Fawn Lily/Trout Lily (*E. dens-canis*), and *Avalanche Lily (*E. grandiflorum*)**
Cooked bulbs are edible. *E. oregonum* is poisonous.

***Dropwort (*Filipendula vulgaris*)**
Leaves were once used in soups and salads.

**Evening Primrose (*Oenothera biennis*)**
The roots of first year plants can be cooked; peppery spring shoots and leaves are used in salads.

***Fennel Flower/Black Cumin (*Nigella sativa*) and *Love-in-a-mist (*N. damascena*)**
Fennel flower's spicy seeds are used in Russia and the Near East to flavor breads and cakes; seeds of love-in-a-mist taste milder and sweet.

***Flowering Rush (*Butomis umbellatus*) (See figure 3-87.)**
Tubers that form along the roots can be roasted.

**Groundnut (*Apios americana*)**
The tubers were boiled, roasted, and fried by American Indians and early settlers.

***Hollyhock (*Alcea rosea*)**
Leaves were once cooked as potherbs; roots yield a nutritious starch.

**Ice Plant (*Mesembryanthemum crystallinum*)**
Leaves can be eaten raw in salads or cooked as potherbs.

**Jicama (*Pachyrhizus tuberosus*)**
The crisp-fleshed roots are edible raw or cooked. Seeds are poisonous.

***Lavender, English (*Lavandula angustifolia = L. officinalis*)**
Leaves, flower petals, and buds were once used as seasoning in Europe; the flowers can be made into conserve.

***Lotus, American (*Nelumbo lutea*)**
American Indians ate the seeds; roots are probably not edible.

***Lotus, Sacred/East Indian (*Nelumbo nucifera*)**
Leaves, flower stems, flower petals, seeds, and young roots are eaten in China. Sacred lotus is one of the most beautiful of all cultivated plants.

**Luffa/Sponge Gourd (*Luffa aegyptiaca*)**
Young cylindrical fruits are used like squash; dried insides of mature fruits are used as sponges; some varieties of dishcloth gourd, *L. acutangula,* with angular fruits may be poisonous.

**Malabar Spinach (*Basella alba*)**
Leaves are cooked as potherbs.

**Marigold, Sweet-scented/Sweet Mace (*Tagetes lucida*)**
Tarragonlike leaves are used as seasoning in Mexico and Central America.

**Marshmallow (*Althaea officinalis*)**
Roots add gumminess to other foods; leaves were once cooked as potherbs.

**Miner's Lettuce/Winter Purslane (*Montia perfoliata*) (See figure 3-88.)**
The succulent leaves are used raw in salads.

**Mitsuba/Japanese Parsley/Honewort (*Cryptotaenia canadensis*)**
In Japan, young leaves and leafstalks are eaten raw or cooked.

**Orach (*Atriplex hortensis*)**
Leaves are cooked as potherbs.

**Oxalis, several species: *Common Wood Sorrel (*Oxalis acetosella*), *Lucky Clover/Rosette Oxalis (*O. deppei*), *Sourgrass/Bermuda Buttercup (*O. pes-caprae*), and *Violet Wood Sorrel (*O. violacea*)**
Leaves of all four can be added sparingly to salads, potherbs and sauces; they contain oxalic acid so they should be used only in small quantities; cooked bulbs of *O. pes-caprae* and *O. violacea* are edible; creeping sorrel *(O. corniculata)* may be poisonous.

**Peanut (*Arachis hypogaea*)**
Nuts are edible.

**Perilla/Beefsteak Plant (*Perilla frutescens*)**
In Japan, the leaves, seedlings, flowers, and seeds are used for seasoning.



***Pond Lily, Common Yellow/Spatterdock (*Nuphar advena*)**
Cooked roots are edible; seeds can be ground, boiled, or popped like popcorn.

3-88. Miner's lettuce. (From Mm. Vilmorin-Andrieux's *The Vegetable Garden.*)

***Pond Lily, European (*Nuphar luteum*)**
Roots can be boiled, roasted, or fried.

***Pond Lily, Western (*Nuphar polysepalum*)**
Cooked seeds are edible.

***Redbud/Judas Tree (*Cercis canadensis* and *C. siliquastrum*)**
The juicy, sour flowers are used raw in salads, pickled, or cooked.

**Red Valerian (*Centranthus ruber*)**
Leaves can be eaten raw in salads or boiled as potherbs.

***Roselle/Jamaican Sorrel (*Hibiscus sabdariffa*)**
Sour-tasting flower sepals and bracts are used in preserves, drinks, and sauces.

***Saffron (*Crocus sativus*)**
Flower stigmas are used dried as seasoning. Do not confuse the similar, but poisonous, "fall crocus" *(Colchium autumnale).*

**Sea Kale (*Crambe maritima*)**
The leafstalks are blanched and harvested in spring during the plants' second year of growth. They can be eaten raw or cooked.

**Sedum, several species: Wormgrass (*Sedum album*), *S. reflexum, S. rosea,* Orpine/Live-forever (*S. telephium*)**
The succulent leaves can be added to salads and soups; *S. acre* is poisonous.

**Serpent Cucumber/Edible Snake Gourd (*Trichosanthes anguina*)**
Immature fruits are sliced and cooked in Asia and Africa.

***Skirret (*Sium sisarum*)**
Cooked roots are sweet-tasting.

***Sorghum (*Sorghum bicolor* = *S. vulgare*) (See figure 3-89.)**
Milletlike seeds of some sorghums are cooked in breads and cereals in Africa; stems of others are crushed for sweet syrup.

***Sweet Cicely (*Myrrhis odorata*)**
Leaves and seeds are used as seasoning; cooked stems and roots are edible.



***Sweet Potato (*Ipomoea batatas*)**
Cooked tubers are edible.

**3-89.** Sweet sorghum.

***Taro/Dasheen (*Colocasia esculenta*)**
Skinned, *thoroughly cooked* corms can be eaten like potatoes or made into Hawaiian poi.

***Tiger Lily (*Lilium lancifolium*)**
Bulbs can be boiled; dried flower buds—"golden needles"—are added to Oriental soups.

**Turnip-rooted Chervil (*Chaerophyllum bulbosum*)**
Sweet-tasting roots can be steamed, boiled, or added to stews.

***Unicorn Plant (*Proboscidea louisianica*)**
Young fruits can be pickled.

**Virginia Mountain Mint (*Pycnanthemum virginianum*)**
American Indians seasoned soup and meat with the flower buds and flowers.

**Wasabi/Japanese Horseradish (*Eutrema wasabi* = *Wasabi japonica*)**
Rhizomes and leafstalks are used as spicy Japanese condiments.

**Watercress (*Nasturtium officinale*)**
Leaves can be used raw in salads and sandwiches, or cooked in soups and stews.

**Watershield (*Brasenia schreberi*)**
In Japan the young leaves are eaten raw or added to soups; American Indians cooked the roots.



***Wild Ginger (*Asarum canadense*)**
Roots can be dried and used as seasoning.

**Yams, two species: *Dioscorea alata* and *D. batatas***
Roots can be cooked like potatoes.

**Yard-long Bean/Asparagus Bean (*Vigna unguiculata* subspecies *sesquipedalis*)**
Pods are cooked when young and tender.

***Yautia/Yellow Yautia (*Xanthosoma sagittifolium*) and *Yautia/Violet-stemmed Taro (*X. violaceum*)**
Skinned *thoroughly cooked* roots are edible.

**Yarrow (*Achillea millefolium*)**
In Europe, the leaves were once used in salads and soups.

Many grains are attractive, and some—rice, wild rice, and oats—are especially good-looking. They can be raised in small patches, or in fields shared by community gardeners or owned or rented by individuals. *Small-scale Grain Raising* by Gene Logsdon (see Bibliography) discusses these possibilities in detail.

Edible fish can thrive in ponds with aquatic food plants; small birds and mammals can live among fruit and nut trees. *Permaculture One* by Bill Mollison and David Holmgren and *Permaculture Two* by Bill Mollison (see Bibliography) describe the ways in which permanent edible landscapes can be the basis for a whole ecosystem and way of life.

## Chapter 4

# Fruits and Nuts for the Beautiful Food Garden

Most fruits and nuts are perennial, and many are large and long-lived. They give the garden a strong, permanent structure that smaller, shorter-lived vegetables and herbs cannot provide. Many fruits have showy flowers and fruits; many nut trees have attractive foliage, impressive size, and distinctive bark and branching patterns.

Growth habits vary greatly and can be utilized in many ways in the landscape. For example, you might use spreading trees for shade, while using those with narrow tops as specimen trees, alone or in groups, in a lawn or garden background. Some trees can be planted close together to form fruiting hedgerows—fences around the garden, or space dividers within it. Large-fruited trees can be trained into productive espaliers on walls, fences, and trellises. You can train dwarf trees, which usually need support, in hedgerows and espaliers or plant them in containers and raised beds.

Plant fruit bushes in hedges or have them share border space with other shrubs. They can grow in raised beds and barrels, and their rounded shapes and regular leaf patterns form an attractive background for flowers, vegetables, and herbs. Arching cane fruits can be used in some of these same ways, and trailing cane fruits can be trained on trellises and arbors.



If you wish, neat, bushy ground cover fruits can be planted in masses, or in rows as edging plants. Some native American woodland fruits trail along the ground and make loose-textured ground covers for shady places.

Keep in mind that fruit- and nut-producing plants are a major commitment. You need to know a great deal about each plant before you put it in your garden, because it will, you hope, grow there for a long time. Which varieties taste best and are most highly recommended for your area? What kind of soil and climatic conditions do they need? What are their ornamental qualities? How much space will the branches and roots occupy? (If possible, look at mature plants of a given variety.) How long will plants take to come into bearing? (Dwarfs generally fruit earlier than do standard-sized trees.) How much maintenance will plants require in terms of pruning and insect and disease control? How much food do they produce and how much of that can your family use? If you cannot use all of it, what arrangements can you make for sharing, trading, or selling the excess? (Many small fruits are eaten by birds, but nut husks and unharvested large fruits can make a slippery, smelly clean-up problem.)

Almost all commonly grown fruits, and nuts (to a lesser extent), have their share of insects and diseases. Many horticulturists recommend extensive spray programs for common fruits, but these may not be necessary if you select locally adapted varieties with good insect and disease resistance and if you will accept less than perfect fruit. Apple worm damage that looks devastating to a commercial grower may be only a minor annoyance to a family that makes most of its apples into applesauce and cider.

Most fruits and nuts need careful pruning in early years to establish a basic branching structure. After that, pruning needs vary from heavy, annual pruning to occasional removal of old, weak branches, depending on the plant.

Spring frosts are crop hazards for early-flowering fruits and nuts. Plants in southern exposures may bloom too early. Avoid frost pockets. Cover smaller plants with a blanket if frost threatens during or after bloom. In areas with vacillating spring temperatures, grow plants that bloom late. If you grow early-flowering plants, you must expect that some years they will produce only foliage.

Since they have specific pollination requirements (described in entries on individual plants), many fruit and nut varieties produce little fruit if grown alone. Others are self-fruitful and produce plenty of fruit when grown alone.

The first several years after a fruit or nut-bearing plant is planted, it needs careful weeding and watering, as well as a mulch to protect its roots. After this, deep-rooted plants can hold their own against competition from smaller plants growing around them. (Crops may be reduced, but you may prefer medium-sized crops to huge ones.)

Fruit trees with open foliage and/or narrow growth habits make more suitable canopy plants than do broad, dense-leaved ones. Spring-flowering bulbs and spring vegetables are very attractive beneath a fruit tree's flowering branches, and they have completed a good deal of their annual growth by the time the fruit tree's leaves block out their sunlight. You can also grow shade-tolerant vegetables, herbs, and flowers under fruit trees. Perennials are especially good because their root systems stay in place, causing less disturbance to tree roots. Wild grasses and flowers can grow together under fruit trees, the way they do in old orchards.

When plants are grown in hedgerows, they need special care, such as pruning, to prevent the most vigorous trees from crowding out the others. Hedgerows can provide a compact food source that doubles as a wildlife refuge, as European hedgerows have been doing for centuries.

Fruits and nuts can be grown in public or semi-public areas, such as streets, neighborhood parks, community gardens, apartment complexes, and housing projects—any place where the crop can be shared by a fairly stable community and where plants are visible from neighbors' windows (so vandalism is kept to a minimum). Fencing may be necessary, but it is better protection to have a large number of neighborhood people involved from the beginning in choosing, maintaining, and sharing the plants. If you use pesticides on plants in a public place, put up warning signs. If possible, place plants with plenty of room to grow, well away from houses, paths, sidewalks, and streets, so that falling fruit does not disturb people.

Be sure to check your local ordinances to make sure fruit and nut tree planting is allowed along the streets.

Common and uncommon fruits and nuts are described in this chapter—both cultivated plants, mostly European and Asian in origin, and wild and partly domesticated American natives. There are various reasons why certain fruits and nuts are widely grown. For one, they may be commercially important because their fruit is attractive and easy to ship. A second factor is that they may be native to areas where people have bred fruit for many centuries, so they are vastly improved from wild forms. (They are not necessarily easy to care for!) Rarely grown plants may produce fruit that has never become popular or that is too soft to ship. The fruit may be good for preserves and juices, but too tart for eating raw or too small to be an important food source. The plants may be of limited cold-hardiness too. Many native American plants are infrequently cultivated, and few varieties exist. They are well adapted and may be very healthy in their native areas but do not always fruit reliably in garden situations. Sometimes little is known about pollination and environmental conditions needed to produce crops. Their culture is still in the experimental stages.

I have not personally grown many of the plants in this chapter. For the information it includes, I am indebted to helpful fruit specialists at Oregon State University—Dr. Melvin Westwood, Dr. Maxine Thompson, and Dr. Alfred Roberts—and to several excellent references: Stanley Schuler's *Gardens Are for Eating,* Donald Wyman's *Wyman's Gardening Encyclopedia,* and Ken and Pat Kraft's *Fruits for the Home Garden.* (See Bibliography.)

Unless otherwise noted, the plants in this chapter are deciduous, and hardy throughout most of the United States, to winter temperatures of −10° to − 20°F. (− 23° to − 29°C.) or lower. Many need a period of winter chilling for breaking dormancy in spring and do not thrive in warm-winter areas. There are strong species and variety differences in cold-hardiness and winter chilling requirements. Local information is essential.

The vegetables and herbs in Chapter 3 were described as simply "hardy" or "tender"—capable or incapable of surviving frost. Because the fruits and nuts in this chapter are long-lived, more information has been gathered on exactly how much cold they can tolerate (and how much winter chilling they require). Cold-hardiness for these plants is described in terms of plant-hardiness zones, which have been developed by the Agricultural Research Service of the United States Department of Agriculture. The approximate range of average annual minimum temperatures for each plant hardiness zone is as follows:

|  |  |  |
|---|---|---|
| Zone 1 | Below −50° F. | (−46° C.) |
| Zone 2 | −50° to −40° F. | (−46° to −40° C.) |
| Zone 3 | −40° to −30° F. | (−40° to −34° C.) |
| Zone 4 | −30° to −20° F. | (−34° to −29° C.) |
| Zone 5 | −20° to −10° F. | (−29° to −23° C.) |
| Zone 6 | −10° to 0° F. | (−23° to −18° C.) |
| Zone 7 | 0° to 10° F. | (−18° to −12° C.) |
| Zone 8 | 10° to 20° F. | (−12° to − 7° C.) |
| Zone 9 | 20° to 30° F. | (− 7° to − 1° C.) |
| Zone 10 | 30° to 40° F. | (− 1° to 4° C.) |

4-1. Delicate flowers of 'Santa Rosa' plum bloom on bare branches.

Unless otherwise noted in individual descriptions, fruits and nuts need fertile, well-drained soil and full sun.

When choosing plants well adapted to your area, try to get information from both professional and amateur sources. Specialists will know the varieties that have insect and disease resistance developed especially for your part of the country. Home gardeners may have plant maintenance styles similar to yours. Fruit trees grown with different care and different amounts of pesticides and fertilizers may behave very differently.

## FRUIT TREES

In spring most fruit trees have beautiful, short-lived, single flowers. (See figure 4-1.) (Fluffy double flowers last longer, but are seldom followed by fruit.) Most fruit trees have plain green, ovate leaves. If the leaves are healthy, they make an attractive background for colorful fruit. They seldom become bright-colored in the fall, however. Bark and branching patterns are pleasant-looking, but not spectacular. Many trees become twiggy if they are not frequently pruned.



Heights range from 4 to 60 feet (1.2 to 18 m). Shapes can be controlled by pruning. Standard-sized trees should be pruned to branch low for easy harvest and maintenance. Dwarf and semidwarf trees, created genetically or by grafting standard stock on dwarfing rootstocks, are often weak-rooted and need support. You can grow them in containers on a sunny patio to take advantage of maximum light. Be sure to give special attention to drainage, pruning, watering, and fertilizing, however.

Most fruit trees require bees for good pollination, and bees are most active when weather is warm and sunny.

In home gardens the most commonly grown fruit trees are apples, pears, plums, cherries, and peaches.

## Apple (*Malus pumila* and crosses with other species: *M. sylvestris, M. baccata,* etc.)

**Zones:** 3 to 8; a few varieties can be grown in Zone 9.
**Growth:** Dwarfs grow 8 to 12 feet (2.4 to 3.6 m) tall; standard trees grow 25 to 40 feet (7.5 to 12 m) tall, and spread about 25 feet (7.5 m).
**Characteristics:** Apple trees begin bearing at four to seven years, dwarf trees earlier. The trees are very productive for twenty to twenty-five years. Some bear on an alternate-year basis. Most apples require cross-pollination to develop fruit. A few, notably 'Golden Delicious,' are self-fruitful, but set larger crops when other varieties are present.

Apple trees have gray, mottled bark. Their flowers are white with a pink blush and bloom after the ovate leaves unfold. Leaves are bright light green at first and later fade to dull green. Insects and diseases often disfigure them by the time the fruits turn color. Fruits are yellow, red, or an inconspicuous green, ripening from July to October depending on variety. Trees need annual pruning and can be trained into open, sculptural shapes.

Apple trees are subject to a large number of insects and diseases, and no variety is resistant to them all. Insects include codling moth, aphids, leaf rollers, maggots, bark borers, and mites. Fruits are prone to several physiological disorders. Apple scab, powdery mildew, fire blight, and apple cedar rust are important diseases. Some resistance to apple scab is available.

Some authorities claim that it is impossible to produce edible apples without pesticides. Most recommend four to nine applications of a multipurpose fruit tree spray at specific times during the growing season. In my opinion, it is possible to grow a good crop for home use in some areas without spraying. However, the chances of attractive foliage without spray protection are slim. Organic sources suggest oil spray while the tree is dormant and removal of fallen leaves and fruit from under trees.

## Pear (*Pyrus communis*)

**Zones:** 5 to 8 are best; some varieties can be grown in Zones 3, 4, or 9.
**Growth:** Pear trees grow up to 40 feet (12 m) tall; dwarfs available.
**Characteristics:** The tree begins bearing at four to five years and is very productive for twenty to twenty-five years. Though many pears are partially self-fruitful, all need cross-pollination to produce a good crop. Some varieties are not good pollinators for each other.

Compared to apples, pears are easy to grow and to keep in pleasant-looking condition. Small, white flowers bloom as the leaves unfold. Shiny green leaves are broadly ovate and often remain attractive throughout summer. Fruits are usually green or brown and inconspicuous, but there are a few red-fruited varieties. Branches are often stiffly upright when plants are young. Mature trees form pyramidal or oval canopies, not spreading enough for good shade. They are attractive specimen trees, alone or in groups, and the flowers are striking against a background of evergreen trees or hedges. Dwarfs are relatively strong-rooted and can be grown without support.

Pears are less fussy about drainage than are many other fruit trees, and tolerate fairly heavy soils. They bloom about one week earlier than apples do, so their flowers are more subject to frost damage. Mature trees need infrequent pruning. Although pears need nitrogen, they do not need very much of it. If pear trees get too much nitrogen they grow very vigorously and become susceptible to fire blight, their major disease problem. Fire blight resistance is available, notably in 'Seckel,' 'Keiffer,' and many recently developed varieties. Scab, another disease possibility, is less serious than it is in apples. Insects include pear psylla, codling moth, and several others.

If trees are unsprayed, many fruits may still be edible, but most authorities recommend the same spray program for pears as they do for apples. Some pear varieties ripen fruit well on the tree; with others, fruit should be ripened off the tree at cool temperatures.

A range of possibilities opens up as blight-resistant Oriental pears, with small, round, crisp-fleshed fruit—*P. pyrifolia* and *P. ussuriensis* varieties—become available in this country. *P. ussuriensis* has excellent cold-hardiness but blooms especially early. Flower buds are pink and leaves turn red in fall. Its growth habit may be more spreading than that of common pears. 'Ya-li' is a good-tasting variety. *P. pyrifolia,* less hardy than *P. ussuriensis,* has coarsely toothed, leathery, ovate leaves that turn reddish purple in fall. Many varieties have been developed from this species; 'Twentieth Century' is excellent. 'Bradford,' a sturdy *P. calleryana* variety, tolerates tough urban conditions and bears tiny fruits that birds like.

## Plums: European plums (*Prunus domestica, P. insititia,* and *P. cerasifera*); Japanese plum (*P. salicina*); American native plums; hybrids between native plums and other species, especially *P. salicina*

**Zones:** European plums: Zones 5 to 7. Japanese plums: Zones 5 to 9. American plums: Zones 3 to 7.
**Growth:** Plum trees grow up to 25 feet (7.5 m) tall; dwarfs and semidwarfs are available.
**Characteristics:** They begin bearing at four to six years of age. Some European plums are self-fruitful, but most plums require cross-pollination by other varieties of their species or by hybrids with their species.

One or another plum species or hybrid can be grown in almost every part of the United States. In some areas, only a few varieties will grow; in other places, there is a wide range of choices.

Plum's white flowers bloom in early spring and are sometimes damaged by frost. Leaves are medium green, broad or narrow ovate with toothed edges; those of Japanese plum are glossy. The fruits are round or oval, with skins of purple, blue, red, yellow, or green, and flesh that may be firm or melting, tart or sweet. Most plum trees have upright branches, though some are spreading.

Plums tolerate heavy soil better than cherries or peaches, but do not prefer it. Fruits may need to be thinned, though natural "June drop" takes care of much excess fruit. European plums need infrequent pruning, but Japanese plums require a great deal. Plum curculio and brown rot cause real trouble in some years (brown rot is encouraged by rainy, humid conditions), but insects and diseases are not usually severe. Borers, San Jose scale, aphids, black knot, and leaf spot are sometimes problems.

The species *P. domestica* includes prune-plums, green gage plums, and lombards. Prune-plums are firm-textured and sweet, usually blue-skinned. 'Stanley' is a sturdy variety of prune-plum that grows well in several areas of the United States. Green gage plums are high-quality dessert plums. Lombards produce reliable crops of large, medium-quality fruits.

*P. insititia* trees are small and floriferous and produce small fruits—good-tasting mirabelles, and sour damsons used for preserves. Damsons thrive with little care.

*P. cerasifera* includes purple-leaved, early-flowering varieties such as 'Pissardii' or 'Atropurpurea,' 'Thundercloud,' and 'Vesuvius,' which are grown as ornamentals in mild climates. Fruits are edible, but often destroyed by spring frosts.

Japanese plum *(P. salicina)* blooms earlier than does *P. domestica,* and many of its varieties are less cold-hardy. Large, juicy fruits are often red-skinned. It grows very quickly and bears flowers and fruits all along its branches. 'Santa Rosa' and 'Burbank' are widely adapted varieties. Some very hardy hybrids have been developed between Japanese plum and native American plums.

Native plum trees usually have small tart fruits that are best used in pies, sauces, and preserves. Varieties of many species have been selected and are available locally. Among the more important native plum trees are *P. nigra, P. americana,* and *P. munsoniana.*

## Sweet cherry *(Prunus avium)*; sour cherry *(P. cerasus)*; Duke cherries (*P.* x *effusus*) are sweet cherry–sour cherry hybrids which more closely resemble sweet cherry

**Zones:** Sweet, Zones 6 and 7, and sometimes 5. Sour, Zones 4 to 7.
**Growth:** The sweet cherry tree grows up to 60 feet (18 m) tall; dwarfs and semidwarfs are available. Sour cherry trees grow up to 35 feet (10.5 m) tall; dwarfs and semidwarfs are available, but dwarfs may produce a small crop.
**Characteristics:** Sweet cherry trees begin bearing at five to seven years of age. Sweet and Duke cherries are not self-fruitful and varieties must be carefully chosen because not all have compatible pollen. Sour cherry trees begin bearing at four to five years of age. They are very productive for fifteen to eighteen years and are self-fruitful.

Cherries' prolific white flowers bloom as the leaves emerge. Leaves are medium green, lanceolate with toothed edges. Sour cherry leaves are smaller and more pointed than those of sweet cherry. The glossy red, yellow, or dark purple fruits shine against plain green leaves. Red sour cherries are translucent and look as if they are made of glass. After you harvest the fruit, the cherries are plain foliage plants for the rest of the summer. They sometimes turn subdued yellow or orange in the fall. Young cherry bark is satiny brown, split into strips by horizontal lenticels. It shows up well in winter, as do the spiky-looking buds that stud the branches.

Sour cherries have spreading branches; sweet cherries are more upright. Although cherries can grow tall, they are easier to care for and harvest if kept fairly short. Sour cherries are generally easier to grow than are sweet or Duke cherries. Sweet cherries need very well-drained soil. They require full sun to crop well, but sour cherries may adapt themselves to partial shade. Sour cherries bloom later than sweet cherries do; their flowers are seldom damaged by frost. Rain causes cherry fruits to crack.

Possible insects include cherry fruit fly maggots, plum curculio, cherry sawfly larvae, black cherry aphids, and tent caterpillars. Possible diseases are brown rot, bacterial canker, leaf spots, black knot, and viruses. Birds can be a problem, especially with sweet cherries. Small trees can be covered with nets for protection.

Different sweet and Duke cherry varieties do well in different parts of the country. Sour 'Montmorency' is well adapted to a range of localities, and, when fully ripe, fruits are sweet enough to eat right off the tree. Black or rum cherry *(P. serotina)* is an attractive native American cherry, with big racemes of fragrant white flowers, long leaves on drooping branches, and small fruits.

## Peach (*Prunus persica*)

**Zones:** 6 to 8 are best; some varieties can be grown in Zones 5 or 9.
**Growth:** Peach trees grow up to 25 feet (7.5 m) tall; dwarfs and semidwarfs are available.
**Characteristics:** Peach trees begin bearing at two to four years of age and stay very productive for ten to twenty-five years. Most varieties are self-fruitful.

Peaches burst into prolific bloom in early spring. Pink flowers are often large and very showy, with single or semidouble petals. Shiny green, lanceolate leaves have finely toothed edges. In late summer, fuzzy skinned fruits ripen and turn yellow, orange, and red in glowing contrast to the foliage. Peaches grow quickly into small, spreading trees; dwarfs grown in containers become round bushes.

Peaches bloom early and are very subject to spring frosts. They do not tolerate poorly drained soil. They set fruit heavily and require fruit thinning; peaches taste best when tree-ripened in warm or hot summers. After the first five years of life, standard peaches need heavy annual pruning to stimulate new growth. Many dwarfs require little or no pruning. Peaches are attacked by insects and diseases that are not only disfiguring but are also potentially fatal and, for this reason, they are dependent on extensive spray programs to remain healthy. Insects and diseases include plum curculio, Oriental fruit moth worms, borers, aphids, mites, scale, brown rot, bacterial leaf spot, and peach leaf curl.

'Redhaven' and 'Reliance' are widely adapted varieties with good cold-hardiness. Nectarines are smooth-skinned peaches, grown mostly in California. Most are highly susceptible to brown rot, but 'Morton' may be resistant. Several peaches bred for ornamental use also produce medium-quality fruit. Examples are 'Saturn' and 'Double Delight,' with extra-large pink flowers, and purple-leaved 'Atropurpurea.' *P. davidiana,* David's peach, flowers very early, and its small, edible fruits are usually destroyed by frost.

4-2. Quince.

## LESS COMMONLY GROWN FRUIT TREES

Some fruit trees that are seldom grown for food are nevertheless thoroughly domesticated and available from nurseries.

Many crabapples (*Malus* species and hybrids) have good-sized fruits, 1 to 2 inches (3 to 5 cm) in diameter. Some taste sweeter than others. The culture and appearance of crabapple trees are similar to those of apple trees, but the fruits are smaller and more plentiful and the flowers are sometimes more prolific. Some crabapple varieties bloom and bear profusely only in alternate years. All are susceptible to scab and borers and some are susceptible to fire blight. Crabapples most often recommended are 'Dolgo,' 'Hyslop,' 'John Downie,' and *M. baccata* var. *mandschurica*. Others include 'Whitney,' 'Cardinal,' 'Transcendent,' and *M. coronaria* and *M. baccata* varieties.

Quince *(Cydonia oblonga)* bears large white flowers in late spring, when soft, broad green leaves have already unfolded. (See figure 4-2.) Gray, mottled branches are low and spreading. The trees reach a maximum height of about 25 feet (7.5 m). Quinces are self-fruitful and each year bear abundant crops of large, bumpy, yellow fruits with tart, aromatic flavor. Fire blight, wormy fruit, and other problems are possible. 'Orange,' 'Pineapple,' 'Smyrna,' and 'Portugal' are the most common varieties (Also, see flowering quince, on page 222).



Oriental persimmon *(Diospyros kaki)* is a round-topped shade tree up to 40 feet (12 m) tall. It is occasionally espaliered. Broad green leaves turn purplish in fall. Flowers are inconspicuous, but the large fruits are flaming orange and may cling to the branches after leaf fall.

Though male and female flowers are on separate trees, females of some varieties set fruit parthenocarpically (without pollination). 'Fuyu' usually does this and has added advantages of extra cold-hardiness and nonastringent fruit. Persimmon trees are hardy to Zone 7 and need long, warm summers to ripen fruit well. They have few insect and disease susceptibilities. Oriental persimmons should be planted in a permanent position early so their taproots are not injured, and trained to a sturdy branching pattern that can bear heavy fruit loads.

Native American persimmon *(Diospyros virginiana)* grows 50 feet (15 m) or taller and bears small orange fruits. Bark is deeply fissured into symmetrical blocks. Culture is similar to that of Oriental persimmon, but plants are hardy to Zone 5. Male and female trees are necessary for fruit set. 'Garretson' is a fine nonastringent variety.

Black mulberry *(Morus nigra)* grows to 30 feet (9 m) and has crooked, spreading branches. Its dark green leaves are deeply toothed and sometimes lobed. Inconspicuous flowers are followed by soft, juicy, purple fruits, which cover the ground as they fall. It is hardy to Zone 6. 'Persian' is a variety worth trial. Red mulberry *(M. rubra)* is similar; 'Travis' and 'Stubbs' are worth trial. White mulberry *(M. alba)* has attractive foliage; silkworms feed on its foliage, but fruits are usually insipid.

Fig *(Ficus carica)* has large, dark green leaves with deep, rounded lobes, borne on thick, smooth branches. It grows to 20 feet (6 m) and may be trained as a tree or shrub. Its flowers are not visible, and fruits are brown or purplish, produced and ripened best during long, warm summers. Some varieties are self-fruitful. Among these are 'Adriatic,' 'Brown Turkey,' 'Celeste,' 'Desert King,' and 'Latterula.' They are hardy to Zone 8 and may survive in lower temperatures with winter protection.

Apricot *(Prunus armeniaca)* does well in some areas, especially on the west coast of the United States, but blooms so early that flowers and young fruits often are killed by frost. It is closely related to peaches, and the culture is similar. It grows to 30 feet (9 m), has beautiful white or pinkish flowers, and has foliage that sometimes turns red in the fall. It is hardy to Zone 5 or sometimes Zone 4. 'Moorpark' blooms somewhat later than other varieties; new varieties have recently been developed for eastern and midwestern parts of the United States.

## UNCOMMONLY GROWN FRUIT TREES



The trees that follow are rarely grown for fruit, although some are commonly used as ornamentals. Heights listed are maximums for mature trees. Unless otherwise noted, all are known or believed to be self-fruitful. (Since some are wild plants, little is known about pollination.) Most have small, tart fruits best used in preserves, sauces, juices, and wines.

In addition to the following list, a large number of attractive trees that produce delicious fruits grow in subtropical areas, Zones 9 and 10; many are evergreen. They include olive, avocado, date, pomegranate, loquat, jujube, kumquat, and *Citrus* species—orange, tangerine, tangelo, sour orange, grapefruit, and lemon.

Some species of hawthorn *(Crataegus)* have fruits large and tasty enough to be used like crabapples. Most have dense, thorny branches. They thrive in a wide range of soils, but are attacked by many insects and diseases, one of the most serious of which is fire blight. Downy hawthorn *(C. mollis)*, 30 feet (9 m) tall, has red and white flowers, large leaves, and sweet, 1-inch (3-cm) fruits. Chinese hawthorn *(C. pinnatifida* var. *major)*, 18 feet (5.4 m), has similar flowers and fruits, and shiny, deeply lobed leaves with good fall color. Thorns are short or entirely absent. It is hardy to Zone 6. Kansas hawthorn *(C. coccinoides)* 20 feet (6 m), has large white flower clusters, small, dark red fruits, and leaves that are red when they first emerge in spring, and turn red again in fall. Mayhaw *(C. aestivalis)* bears fruits in late spring, and *C. azarolus* has apple-flavored fruit. The latter two species are hardy to Zone 7.

Elderberries (several *Sambucus* species) have broad, cream-colored flower clusters, large, compound leaves, and big bunches of fruit. They grow quickly and break easily. Their branching, if unpruned, often looks awkward, especially during winter. They thrive in a wide range of soils and climates, in sun or partial shade. Two native American species, American elder *(S. canadensis),* 12 feet (3.6 m), and blueberry elder *(S. caerulea),* 45 feet (13.5 m), have seedy, dark blue fruits. *S. canadensis* varieties include large-fruited 'Adams' and 'York,' productive 'Johns,' and mild-flavored 'Superb.' 'Acutiloba' has fine-textured leaves and 'Aurea' has yellow leaves and red fruits. European elder *(S. nigra)*, 30 feet (9 m), has shiny black berries; varieties include 'Aurea and 'Laciniata' ("cut-leaf"). Blueberry elder and European elder are hardy to Zone 6, and American elder to Zone 3. Many parts of elders are poisonous. Some sources recommend eating flowers, fried as fritters or pickled as capers, but others say the flowers cause nausea. Fruits are safe to eat after they have been cooked, but seeds should be removed.

Serviceberry or shadbush (native *Amelanchier* species) have smooth, gray bark, and white flower clusters in early spring. (See figure 4-3.) Leaves are coppery in spring and turn red, orange, or yellow in fall. Small, juicy fruits ripen in early summer. Plants grow in sun or light shade, but do not always fruit heavily in either location. Serious insect and disease possibilities are lacewing fly, red spider, various scales, and fire blight. Shadblow serviceberry *(A. canadensis)* is an up-

4-3. Serviceberry in flower.

right tree, 20 to 60 feet (6 to 18 m) tall, with silvery spring foliage and purple fruits. Allegheny serviceberry *(A. laevis)* is a broader tree, to 35 feet (10.5 m), with red berries. Apple serviceberry *(A.* x *grandiflora),* 25 feet (7.5 m), has large flowers and spreading branches. *A. florida,* 30 feet (9 m), is a native of the West Coast of the United States.

Pawpaw *(Asimina triloba)*, native to the eastern United States, is a slow-growing tree or shrub 10 to 35 feet tall (3 to 10.5 m). Large, dark green leaves droop down and swing gently even in light breezes. The leaves turn yellow in fall. Purple cup-shaped flowers are followed by long, brown, lumpy fruits, the flesh of which tastes sweet and melting—too rich for some people's taste. Some plants have poor-quality fruit. Buy named varieties with proven fruit flavor. Plants are hardy to Zone 6; they grow in moist, fertile soil and in sun or light shade.

## NUT TREES

Most nut trees have inconspicuous flowers that are wind-pollinated and fruits that are not colorful. Their ornamental value lies in their foliage, bark, and growth habit. They often reach impressive dimensions. They are good shade trees, but few can be grown in small yards and no dwarfs are available. They seldom need extensive pruning or spraying, but when they do you may need professional help to cope with large nut trees as you would with any large trees. In many cases, two different varieties must be planted together to produce good crops. Some nut trees take a long time to come into bearing; most continue to bear for a long time. Some crop unreliably because of spring freezes and/or a tendency to biennial bearing. Squirrels are often a problem too. You can rejoice when you harvest a good crop of nuts, but you cannot depend on it.

City regulations notwithstanding, most nut trees are well suited for life on the street. They often branch high and can be easily harvested as the nuts fall. Above the street they have plenty of light and space to grow, although their roots may be cramped at ground level. The wonderful advantage, of course, is that they produce valuable food that neighbors can share.

## English or Persian Walnut (*Juglans regia*)

**Zones:** 5 to 10, depending on variety.
**Growth:** They grow 40 to 100 feet (12 to 30 m) tall.
**Characteristics:** The trees begin bearing at about five years of age. Walnuts are self-fruitful, but each tree has both male and female flowers, which may bloom at different times. Two varieties with complementary flower timing must be planted together.

English walnuts have smooth, pale gray bark and open, spreading branches. Glossy leaves are divided into pinnate leaflets, which unfold early and can be killed by frost, especially in areas where spring temperatures fluctuate. Trees will be unharmed and grow new leaves in summer, but no flowers or fruit will be produced that year. Carpathians, a group of *J. regia* varieties, are more cold-hardy and disease-resistant than others, but leaf out extremely early under some conditions. 'Colby' and 'Hanson' are popular varieties most often grown in the northeastern United States. Non-Carpathian varieties are grown mostly in the West; 'Franquette' leafs out later than most.

Walnuts require very good drainage, although Carpathians are somewhat flexible. Young trees must be carefully trained to develop a strong branching system; occasional thinning is necessary for older trees. Insects include husk maggots, codling moth larvae, and walnut weevils; and diseases include walnut anthracnose, brooming disease, crown rot, and lead blight. Some disease resistance is available. There are cut-leaf and pendulous forms of English walnut, developed for ornamental rather than food plant characteristics.

## Black Walnut (*J. nigra*)

**Zones:** 4 to 8.

**Growth:** They grow up to 150 feet (45 m) tall. Two varieties are necessary for good yield.
**Characteristics:** Trees begin bearing five or six years after planting.

Black walnut has a tall, straight, main trunk of dark and deeply

4-4. Black walnut.

furrowed bark, with open, rugged branches at the top. (See figure 4-4.) Its gaunt beauty adds majesty to a winter garden or streetscape. In summer, its long groups of pinnate leaflets give light shade; they may turn yellow in fall. Grass can grow beneath it, but it secretes a substance that inhibits the growth of nearby fruits and vegetables of some species. Ripe nuts fall with husks still tightly attached. They need to be picked up frequently if they land on paths, sidewalks, or streets. Black walnuts are very good trees to plant *away* from houses and patios. 'Thomas' and 'Stabler' do well together in many areas.

Husks are hard to remove and the nuts inside are thick-shelled and hard to crack; kernels often come out in fragments, but have a rich, pleasant taste that rewards one for the difficulty of extracting them. Train young trees to one main trunk and strong, widespread branches. Little pruning is needed for older trees. Black walnuts are not fussy about drainage and grow well in moist soils. Insects and diseases include black walnut caterpillar, fall webworm, codling moth, lacebug, walnut bunch disease, and crown rot. Varieties bred for ornamental use include 'Deming Purple' and a cut-leaf form.

## Walnut Relatives

Japanese walnut *(J. ailanthifolia),* and especially heartnut (*J. ailanthifolia* var. *cordiformis*), are fine ornamentals with good-tasting nuts. They have spreading branches and grow about 40 feet (12 m) tall. They unfurl lush, green leaflets in early spring. If young shoots are not killed by frost, they bear long male catkins, small, red female flowers, and then long ropes of small, thin-shelled nuts. Trees bear early in life and are adapted to a wide range of soils and climates. They are not reliably self-fruitful. 'Walters,' 'Bates,' and 'Fodermaier' are frequently mentioned varieties.

Butternut *(J. cinerea)* is a spreading tree that grows up to 90 feet (27 m) tall and bears thin-shelled nuts soon after it is planted. It is subject to several serious diseases.

Culture for butternuts and Japanese walnuts is similar to that for other *Juglans* species. Both fruit better when two varieties grow together. Neither is long-lived, as nut trees go. Japanese walnuts may live only thirty years. "Butterjaps" are *J. cinerea* x *J. ailanthifolia* hybrids.

## Pecan (*Carya illinoinensis*)

**Zones:** 5 to 9, depending on variety.
**Growth:** Pecan grows up to 100 feet (30 m) tall, sometimes 150 feet (45 m).
**Characteristics:** Male and female flowers do not bloom at the same time, so two well-matched varieties must be planted together. The tree begins bearing fruit at about five years of age. The first substantial crops appear at ten years and it produces well for up to one hundred years.

Pecan is widely grown in the southeastern United States and some varieties have been developed to produce well in the central part of the country. Trees may survive in northeastern or northwestern states, but nuts seldom develop fully without long, warm summers. Pecans have tall, straight, main trunks and spreading branches, inconspicuous flowers, and rows of narrow leaflets that turn golden brown in autumn.

Pecans leaf out in late spring. They avoid spring frosts, but may not have time to mature nuts. They grow quickly. Their many insects and diseases include various case borers, black pecan aphid, pecan weevil, hickory shuckworm, and pecan scab. Several varieties are resistant to scab. Some sources recommend a regular spray program, but, as trees grow large, this becomes complicated. Pecan roots are deep and easily damaged by transplanting. While still small, trees should be moved to a permanent position in deep soil.

## Pecan Relatives: The Hickories

Shagbark hickory *(C. ovata)* and shellbark hickory *(C. laciniosa)* grow farther north than pecan and produce good-tasting, hard-shelled nuts. Shagbark hickory, native to the central and eastern United States, hardy to Zone 4, grows up to 120 feet (36 m) tall, with a cylindrical, open canopy. Its compound leaves have relatively few leaflets and turn gold in fall. The dark bark flakes off in long, shaggy strips. The tree grows slowly. In fact, seedling trees may not bear for fifteen years; grafted ones bear earlier. Roots are easily hurt by transplanting. There are varieties of local importance, and productive crosses between shagbark and other *Carya* species. 'Fairbanks' is one such hybrid.

Shellbark hickory is similar, but its nuts have thicker shells and its bark does not peel. Its hardiness is only slightly greater than that of pecan. Nuts of several other native *Carya* species are occasionally eaten, but they are not of good quality.

## Chinese Chestnut (*Castanea mollissima*)

**Zones:** 5 to 8, sometimes 4.
**Growth:** The tree grows up to 50 feet (15 m) tall.
**Characteristics:** Two or more varieties are needed for good production. Grafted chestnuts begin bearing at four to five years of age, and seedling chestnuts several years later.

The dense-leaved branches of Chinese chestnut spread widely and give good shade, and its narrow, toothed leaves are shiny green. When young, they are reddish and have wooly undersurfaces; in fall they turn yellow or gold brown. Long, conspicuous catkins bloom in early summer; nuts are large and rather sweet. Trees grow best if the surface roots are mulched and left undisturbed. Insects and diseases to which Chinese chestnut is susceptible include chestnut borers, chestnut weevils, various leaf-eating caterpillars and die-back fungus. Chinese chestnut is only slightly affected by the chestnut blight that kills American chestnuts.

Japanese chestnut *(C. crenata)* is also quite resistant to chestnut blight but its nuts are more coarse-textured and bitter than are those of Chinese chestnut.

Chestnuts have been used as staple articles of diet in various parts of the world. You can cook them like potatoes—baked, boiled, mashed, roasted with meat—make them into flour, dry them for winter use, or glaze them with sugar for desserts.

4-5. Filbert catkins blooming in a winter snowstorm.

## Hazelnut or Filbert (*Corylus* species, especially *C. avellana*)

**Zones:** 6 to 9.
**Growth:** Hazelnut grows up to 25 feet (7.5 m) tall.
**Characteristics:** Two well-matched varieties are needed for cross-pollination. However, some are "cross-incompatible"—one variety cannot fertilize the other. The tree begins bearing two to three years after planting; there is erratic production after ten years unless the tree is severely pruned about every third year.

*Corylus* species have less massive growth habits than other nuts. Only one, Turkish filbert *(C. colurna),* is a real tree, with growth up to 60 feet (18 m) tall. The others can be trained into spreading trees, or a clump of a few strong trunks, but if left to themselves, they grow into many-stemmed thickets. Leaves are oval, toothed, and deeply veined. They are usually medium green, but those of *C. maxima* 'Purple Aveline' ('Purpurea') are purplish brown. The nuts, borne in fringed, papery husks, are more conspicuous and attractive than other nuts.

*C. avellana* is the species most often grown in the United States; 'Barcelona' is an important variety. Commercial production is almost entirely in Oregon. Filberts are cold-hardy in other areas, but do not always fruit. They bloom in winter (see figure 4-5) and their conspicuous, yellow catkins are hurt by temperatures below 0° F. (−18° C.). Insect and disease susceptibilities include hazelnut weevil, aphids, filbert worm, lacebug, mites, scale, and, in the eastern United States, filbert blight.

## UNCOMMONLY GROWN NUT TREES

Nuts of the stately American beech *(Fagus grandifolia)* have often been used for food and its young leaves have often been cooked as potherbs.

Good-sized pine nuts are borne inside cones of many *Pinea* species, all needle-bearing evergreens. Swiss stone pine (*P. cembra)*, Korean pine *(P. koraiensis)*, and limber pine *(P. flexilis)* are slow growing and very hardy. Himalayan pine *(P. wallichiana),* a huge tree with gracefully drooping needles, and sweet-seeded *P. lambertiana,* tall and narrow-headed, are both hardy to Zone 6. *P. torreyana, P. coulteri, P. sabiniana,* and *P. cembroides* are southwestern United States natives.

Gingko *(Gingko biloba)* is an eccentric, dignified tree of ancient origin and of great health and vigor. It has stiff branches and bright green, fan-shaped leaves that turn clear yellow in fall. Female trees produce soft, fetid-smelling fruits with delicious seeds inside. These "gingko nuts" are used in Chinese stir-fries.

Almond (*Prunus dulcis* var. *dulcis*) and pistachio *(Pistacia vera)* are attractive trees that bear nuts reliably only in Zone 9.

## BUSH FRUITS

Bush fruits can share border space with other shrubs that are purely ornamental. You can plant them in hedges or in raised beds and barrels. As hedge plants they can define or divide space within the garden. They might be used to line a driveway or path; to serve as foundation plantings; or to emphasize a boundary between a patio and a play area. Bush fruits most commonly grown are blueberries, currants, and gooseberries.

### Blueberry, Highbush (*Vaccinium corymbosum* and crosses with other species)

**Zones:** 5 to 7.
**Growth:** Blueberry bushes grow up to 12 feet (3.6 m) tall and about 6 feet (1.8 m) wide.
**Characteristics:** Two or three varieties are needed for good fruit set. The plants begin bearing at two to three years of age.

Blueberries grow in upright bushes. Pink, bell-shaped flowers bloom in May on slender, reddish branches, and are followed by small, dark green, ovate leaves and dark blue fruits. (See figure 4-6.) In autumn the leaves turn bright red. Plants need acid soil that is high in organic matter; they have shallow, fibrous roots and benefit from mulching. They need occasional light pruning to remove old, weak branches.

4-6. Blueberry.

Blueberry bushes are usually quite healthy, but insect and disease possibilities include stem borers, leaf-chewing insects, bud mites, fruitworms, and leaf- and stem-spotting fungi. If birds go after the berries, cover bushes with netting.

Blueberries are attractive all year. They grow well with their acid-loving relatives—rhododendrons, azaleas, and heather.

Rabbit-eye blueberry *(V. ashei)* Zones 7 to 9, is well adapted to the southeastern United States and tolerates a root-rotting fungus that is troublesome there. Evergreen blueberry, evergreen huckleberry, or box blueberry *(V. ovatum)* has neat, shiny dark green leaves, coppery young shoots, inconspicuous pink flowers, and small, tasty, black fruit. It grows in a graceful bush up to 10 feet (3 m) tall. The bush grows dense in full sun and loosely branched in shade. It is hardy to Zone 7. Several other bush blueberry species are grown in areas where they are native.

## Currant (*Ribes sativum*)

**Zones:** 3 to 8.
**Growth:** Currants grow up to 6 feet (1.8 m) tall and usually 3 to 4 feet (.9 to 1.2 m) wide.
**Characteristics:** They are self-fruitful and begin bearing at three years of age. They remain productive for twenty to fifty years.

Currants grow in rounded bushes of dense, maplelike leaves. Racemes of yellow bell-shaped flowers bloom when bushes are leafed out. They are followed by clusters of shining red or white fruits.

Currants are productive, ornamental additions to shrub borders and tree understories. 'Red Lake' is a widely adapted variety.

Currants tolerate some shade and a wide range of soils. Prune out branches over three years old and new branches that are not upright. Insect and disease susceptibilities include aphids, currant worm, anthracnose, and cane blights. Currants carry white pine blister rust, which kills white pines. For this reason, black currant *(R. nigrum)* is outlawed all over the United States, and it is illegal to plant *any* new currants where white pine is a timber crop. It is not a good idea to plant them close to five-needled pines in any area.

Red-flowered currant *(R. odoratum)* has showy pink or red flowers. Clove-scented currant *(R. sanguineum)* has fragrant, yellow flowers and red fall color; 'Crandall' is a large-fruited variety. Both species *R. odoratum* and *R. sanguineum* have small, bland-flavored fruits.

### Gooseberry (*Ribes hirtellum; R. uva-crispa*)

**Zones:** 3 to 8.
**Growth:** Gooseberry bushes grow up to 4 feet (1.2 m) tall.
**Characteristics:** They are self-fruitful and begin bearing at three years of age. They remain productive for twenty to fifty years. Gooseberries have thorny stems and hawthornlike leaves that are lobed and edged by rounded teeth. Bell-shaped flowers are greenish; fruits are yellow, red, green, or pink, and sometimes hairy.

Culture of gooseberries is similar to that of currant. Gooseberries thrive in partial shade and need good air circulation. Plants are carriers of white pine blister rust.

## UNCOMMONLY GROWN BUSH FRUITS

Most of the bushes in this category have small, tart fruits that you can use in jellies, juices, and wines. Seedless or small-seeded fruits are delicious in jams and sauces, or added to pancakes, muffins, cakes, and cookies. Many can be frozen. Most of the fruits are eaten by birds as well as people, so you may have a race to reach them first. Many are on recommended lists for wildlife plantings.

Many of these bushes are American natives, often woodland plants that can grow in sun or partial shade. You can use them for hedges, shrub borders, foundation plantings, and background positions, or plant them underneath light-canopied trees. Quite a few bear fruit soon after they are planted. Unless otherwise noted, they are considered to be self-fruitful, but pollination requirements are sometimes unknown, and some bear unreliably in garden conditions. Maximum heights are listed.

Spreading bushes of Nanking cherry *(Prunus tomentosa),* 5 to 10 feet (1.5 to 3 m) tall, have red and white flowers and small red fruits. Four-foot (1.2-m) dwarf or Korean bush cherry *(P. japonica)* looks similar. Six-foot (1.8-m) sand cherry or western bush cherry *(P. besseyi)* has gray leaves, white flowers, and small black fruits; 'Hansen's' bush cherry is a 4-foot (1.2-m) form with larger fruits. Six-foot (1.8-m) beach plum *(P. maritima)* thrives in poor, well-drained soil and bears small purple plums. Beach plum is native to the eastern United States and sand cherry is native to the Plains states.

Native highbush cranberry *(Viburnum trilobum),* 10 to 15 feet (3 to 4.5 m) tall, has dense maple-lobed leaves, flat white flower clusters, and bunches of juicy translucent berries that turn from yellow to crimson. Foliage turns red in fall. There are several large-fruited varieties. eties. European *V. opulus* looks very similar, but its berries taste bitter. Wild raisin or nannyberry *(V. lentago)* is a 25-foot (7.5-m) bush or small tree, with lacy clusters of cream-colored flowers, blue-black fruit, and shiny green leaves that turn purplish red in autumn. It grows well in moist soils. Stagbush or black haw *(V. prunifolium)* is similar, but it has smaller leaves and grows better in dry soil; it is often trained into a small tree. Hobblebush *(V. alnifolium)* grows in a fan-shaped bush 3 to 10 feet (.9 to 3 m) tall and does well in moist conditions. Its leaves turn red in fall. Nannyberry, stagbush, and hobblebush have large-seeded fruits that are sweet and edible but often dry. All are native to the eastern United States.

Cornelian cherry *(Cornus mas),* 20 feet (6 m), is a vigorous, spreading bush with small, starlike yellow flowers covering its bare branches in early spring. It has plain, glossy green foliage that turns reddish in fall and small, red fruits. Fruiting is unreliable in some locations and plants grown from seed produce no fruit until they are ten to fifteen years old. Make sure to get cornelian cherries grown from cuttings of mature plants.

Medlar *(Mespilus germanica),* hardy to Zone 6, grows into a bush or small tree 20 feet (6 m) tall. Flowers are white or pinkish, leaves plain and dull green, and fruits brown and lumpy-looking. The fruits are edible only after they have been "bletted" (allowed to freeze and partially rot), at which point they taste pleasantly acid. 'Royal' is a sturdy variety.

Six- to 10-foot (1.8- to 3-m) flowering quince *(Chaenomeles speciosa)* and its 3-foot (.9 m) relative *C. japonica* have showy spring flowers that are red or orange, sometimes white or pink, single or double. Branches are often thorny. Shiny green, ovate leaves are bronze as they first unfold. Fruits are light green and fairly large; 'Arthur Colby' has especially big ones. Cross-pollination is needed for good fruit production.

Buffalo berry *(Shepherdia argentea)* is a thorny shrub up to 18 feet (5.4 m) tall, with narrow, gray leaves, inconspicuous yellow-green flowers, and clusters of sour, dark orange fruits that turn sweet after frost. Russet buffalo berry or soapberry *(S. canadensis)* is similar, except it grows to only 7 feet (2.1 m) and its berries taste bitter. American Indians whipped them with sweetened water to make a frothy dessert. Male and female flowers of buffalo berries are borne on separate plants. Both sexes are needed for fruit set. Both buffalo berry and soapberry are American natives.

Ten-foot (3-m) cherry eleagnus or goumi *(Elaeagnus multiflora),* from China and Japan, has dark green leaves that are silver underneath, fragrant, yellow, fuschialike flowers, and abundant orange fruits.

Several native *Amelanchiers,* serviceberries, grow as shrubs. *A. alnifolia,* 8 feet (2.4 m) tall, has large flowers and berries compared to other species, and *A. stolonifera* spreads into 4-foot (1.2-m) thickets. (For a general description of serviceberries, see "Uncommonly Grown Fruit Trees.")

Roses with large, well-flavored fruits include *Rosa rugosa* and *R. villosa,* which are described in Chapter 3.

Several edible, native *Mahonia* species have erect racemes of bright yellow bell-shaped flowers, purple berries, and dark, lustrous, evergreen foliage that turns red-brown in cold weather. The berries make excellent jelly. Oregon grape *(Mahonia aquifolium)* is an upright shrub 3 to 15 feet (.9 to 4.5 m) tall. Plants reach different sizes depending on genetic makeup and environmental conditions. *M. nervosa* grows only 1½ feet (45 cm) tall, but otherwise resembles *M. aquifolium.* Both grow in full sun, light shade, or deep shade and are hardy to Zone 6.

Evergreen *Gaultheria shallon,* salal, is equally cold-hardy. It is a vigorous, acid-loving plant that grows in sun or shade, from a low ground cover to a 10-foot (3-m) bush, depending on environment. It has racemes of bell-shaped, pinkish flowers, broad, leathery, light green leaves that are carried on pink leaf stalks and that turn reddish in cold weather, and black fruit that tastes sweet and rather dry. It spreads into clumps.

Oso berry or Indian plum *(Oemleria cerasiformis),* 15 feet (4.5 m), has bright green leaves, drooping clusters of white flowers, and small black fruits on female bushes. It is native to the West Coast of the United States and hardy to Zone 6.



Fuchsias described as food plants *(F. corymbiflora, F. denticulata,* and *F. racemosa)* are not commonly cultivated, but many people eat the tart, purplish black fruits of hybrid garden fuchsias without any ill effects. Plants have gracefully arched branches and dense, ovate leaves. Flowers are white, red, pink, purple, and combinations of these colors. They may be single or double, large or small. They do best in cool summers and partly shaded locations. They are frost-tender; in cold-winter climates you can cut them back and move them into a cool, dark room until spring. They do not produce fruit reliably.

## CANE FRUITS

Raspberry and blackberry plants, *Rubus* species, have long, flexible stems, or "canes," that are sometimes upright and arching, sometimes trailing and vinelike. Canes are biennial—they produce fruit their second year and then die—and the perennial crowns send up new canes to replace them.

### Red Raspberry (and occasionally yellow) (*Rubus idaeus*)

**Zones:** 3 to 7.
**Growth:** Red raspberry grows to 6 feet (1.8 m) tall and 6 feet (1.8 m) wide if the suckers are removed.
**Characteristics:** It is self-fruitful and bee-pollinated. It comes into bearing the second year after planting and bears well for five to twelve years.

Raspberries have three-part, light green leaves with deep veins and toothed edges; canes are arching and bristly or thornless. Flowers have small white or pink petals and prominent greenish stamens and sepals; the bright fruits are conspicuous.

Raspberries need good air circulation and full sun or partial shade. They do not thrive in hot summers. Insect and disease susceptibilities are numerous, but you can minimize problems if you plant resistant varieties and certified virus-free stock, if you remove and burn diseased canes and plants, if you avoid soil previously used for other raspberries or for tomatoes and their relatives, and if you weed out nearby wild brambles. A four-time spray program is sometimes suggested. You can grow the plants in hills or hedgerows. They can be loosely supported by wires stretched between stakes, but you do not have to tie them. A raspberry hedge looks good in front of evergreens; bright new leaves stand out against the dark background. You can grow a double row of raspberries behind smaller, showier vegetables and herbs; the arching canes of thornless varieties can form a fruit-bearing hiding place for small children.



Most varieties bear in early summer, but "ever-bearing" varieties have a light crop in July and a heavy one in fall. Varieties vary greatly in fruit flavor, ripening date, disease resistance, and cold-hardiness.

Black raspberries *(Rubus occidentalis)* and purple raspberries *(R. x neglectus)* are larger, more vigorous, and more erect than red ones. They tolerate hot weather better than do red raspberries, but some varieties are more susceptible to disease. They are most often used in preserves.

## Blackberry (also Loganberry, Boysenberry, and others) (*Rubus argutus, R. ursinus, R. laciniatus, R. procerus,* and other species)

**Zones:** Without protection, Zones 5 to 8, or 7 to 8, depending on variety.
**Growth:** Canes grow up to 20 feet (6 m) long.
**Characteristics:** Plants are self-fruitful, but cross-pollination increases fruit size. They come into bearing the second year after planting and remain productive for six to fifteen years.

Blackberry canes are long, arching or trailing, and often very thorny. Leaves are gray-green or yellowish green, toothed and compound, with three to seven leaflets. One variety, 'Thornless Evergreen,' has leaves that last all winter; the leaves are attractive and deeply cut, but the plant is hardy only to about 0° F. (− 18° C.). After clusters of pale pink flowers bloom, glossy fruit clusters turn from green to red and then black.

Blackberry culture is similar to that of raspberry. In cold climates, laying canes down on the ground and mulching them can greatly lessen winter damage. Insect and disease susceptibilities are somewhat less severe than are raspberry's, but the same control measures are necessary. Since plants are big and thorny, prompt pruning and heading back are important. Hoe up the invasive suckers wherever they appear in the garden. Cold-resistant, erect varieties like 'Darrow' are usually grown in the northeastern United States; trailing and thornless varieties with less cold-tolerance are well adapted to the Northwest.

You can grow blackberries with a post and wire system and use them as a garden background. You might trellis them to provide screening and shade or have them silhouetted on a sunny wall. Less rampant varieties can ramble on a fence with climbing roses, grapes, or autumn clematis.

## Other *Rubus* Species

Less common *Rubus* species have fruits with uncertain fresh flavor that are best used in preserves. Many of these species do well in partial shade and grow up to 9 feet (2.7 m) tall. Flowering raspberry *(R. odoratus)* has arching canes, fragrant purple flowers, and red fruits that appear unreliably. Thimbleberry *(R. parviflorus)*, 6 feet (1.8 m) tall, is very erect and has downy, maple-lobed leaves of bright yellow-green; large white flowers; and sweet, seedy, red berries. It is a wild plant of the western United States and does not always thrive under garden conditions. Wineberry *(R. phoenicolasius),* hardy to Zone 6, has graceful, arching canes covered with long red-brown hairs; leaves are three-part, clusters of small flowers are pink or white, and berries are bland. *R. spectabilis* has reddish brown bark, bright pink flowers, and bland fruit. *R. deliciosus,* hardy to Zone 6, has thornless, arching canes, large, white flowers with prominent yellow stamens, and fruit variously described as "delicious" and without flavor."

## VINE FRUITS

Grape, which is the most commonly grown vine fruit, is described in Chapter 3.

Kiwi or New Zealand gooseberry *(Actinidia chinensis)* is hardy to Zone 7; it requires both male and female plants and a long, warm season to produce good fruit. It may quickly cover a large area. It has big, hairy, green leaves that are reddish in spring; pretty white flowers; and brown, hairy-skinned fruits with delicious green flesh. Bower actinidia *(A. arguta)* is equally vigorous, and hardy to Zone 5. It has small white flowers, shiny, deciduous leaves with red leaf stalks, and small, bland fruits that taste good in preserves.

Passion flower *(Passiflora caerulea)* and maypops *(P. incarnata)* have lobed leaves, beautiful flowers, and edible fruits. They are hardy to Zones 7 or 8.

## GROUND-COVER FRUITS

Strawberry is the only ground cover fruit that is commonly grown by home gardeners.

### Strawberry (*Fragaria* hybrids; alpine strawberries *F. vesca*)

**Zones:** 3 to 10, depending on variety.
**Growth:** Herbaceous perennial.
**Characteristics:** They are self-fruitful and come into bearing the same year that they are planted. There are one to three years of heavy crop, although many plants live longer and produce more lightly.

Strawberries grow in low rosettes. The three-part, glossy, dark green leaves have scalloped edges. (See figure 4-7.) They are evergreen in mild climates. The yellow-centered white flowers are often still blooming as fruits turn red. Strawberries need a steady supply of water and soil nutrients as well as freedom from late spring frosts.

4-7. Strawberry.

They have plenty of insect and disease susceptibilities, so you should buy virus-free plants and look for varieties resistant to other diseases that are problems in your area.

Plant strawberry crowns level with the surface of the soil. During the year that they are planted, pinch off flowers so plants can use their energy to establish themselves and to produce runners. Second-year plants bear heavily, and then crops may decrease. There are various methods of renovating strawberry plantings to prevent this. Winter mulching increases cold-hardiness.

There is a large number of strawberry varieties, both June-bearing and ever-bearing, many of which have been developed for specific localities.

Traditionally, strawberries are planted in "hills"—heaped-up soil—or "matted rows"—double rows with room between for runner plants to grow. You can use strawberries as ground covers or edging plants, alone or mixed with low, annual flowers—dwarf phlox and dianthus—spaced freely to avoid root competition. Strawberries in containers, such as pyramids, barrels, jars, and window boxes, are generally productive for only one year.

Wild American strawberries have small, sweet berries. Coast strawberry *(F. chiloensis),* native to the West Coast of North America and Chile, has thick, glossy leaves, and male and female flowers on separate plants. Both sexes are needed for fruit production. *F. virginiana* is native to the eastern United States.

Varieties of runnerless alpine strawberry *(F. vesca),* such as 'Baron Solemacher,' 'Alexandria,' and 'Harzland,' can be grown from seed. Germination is sometimes poor. Started indoors in April, they bear fruit from August until frost. After winter they sprout up again, surrounded by numerous seedlings.

## UNCOMMONLY GROWN GROUND-COVER FRUITS

Many ground cover fruits are North American woodland natives that grow well in partial shade and moist, acid soil rich in organic matter. They can be grown at the base of trees and shrubs. Not all fruit reliably, and little is known about pollination requirements.

Lowbush blueberry *(Vaccinium angustifolium)* grows in 2-foot (.6-m) mats. Its culture is similar to that of highbush blueberry, but it is native to colder regions (Zones 3 to 5). Other low-growing blueberry species include *V. vacillans, V. myrtilloides,* and others.

Cranberry *(Vaccinium macrocarpon)* is native to bogs in northeastern America. (It is planted commercially in bogs.) In these conditions plants are hardy to − 50° F. (− 46° C.), but when grown in drier places their flower buds and stems may be hurt at temperatures well above 0° F. (− 18° C.). Cranberry is a low-growing, fine-textured plant with small, ovate leaves, star-shaped, pink flowers, and red fruits. European cranberry *(V. oxycoccos)* and cowberry or lingonberry *(V. vitis-idaea* and its variety *minus*) also have good fruits, and red leaf color in fall.

Strawberry-raspberry *(R. illecebrosus)*, hardy to Zone 6, has large white flowers and sour red fruit, often both on the plant at the same time. It grows in dense, low, prickly mats.

Oregon wintergreen or mountain checkerberry *(Gaultheria ovatifolia)* looks like a miniature salal plant, but has red berries that taste like wintergreen. It is hardy to Zone 6.

Snowberry *(Gaultheria hispidula)* has trailing stems, shiny green leaves, small white flowers, and white berries that taste like wintergreen. Northeastern American Indians made tea from the leaves.

Box huckleberry *(Gaylussacia brachycera)* grows slowly into mounds about 1½ feet (.5 m) tall. Its evergreen foliage may turn bronze in fall. It is hardy to Zone 6. Black huckleberry *(G. baccata)* 1 to 3 feet (.3 to .9 m) tall, has racemes of pink flowers and oval, deciduous leaves. Plants may spread into thickets. Both species have dark blue seedy berries. *G. frondosa* and *G. dumosa* also have edible fruits.

*Mahonia repens* is an Oregon grape relative that forms a low mat of dull green leaves.

# Appendix: Sources and Information

## VEGETABLE AND HERB VARIETIES

### THE SOURCES

Each seed company and/or nursery listed below is preceded by a letter. This letter will stand for that seed company throughout the vegetable and herb variety list that follows on page 232.

Keep in mind that the information listed here (including catalog prices) was collected in 1982 and is subject to change. Catalog prices are relevant through 1983.

**Major Sources**
The seed companies listed below are those that carry the greatest numbers of vegetable and herb varieties listed in this book.

A Abundant Life Seed Foundation
P. O. Box 772
Port Townsend, WA 98368
($2.00 for 1 year of catalogs and newsletters)

B Burpee Seed Company
Warminster, PA 18991
Clinton, IA 52732
Riverside, CA 92502
(free catalog)

C De Giorgi Company
Council Bluffs, IA 51502
($1.00 for catalog)

D Joseph Harris Company
Moreton Farm
Rochester, NY 14624
(free catalog)

E Nichols Herb and Rare Seeds
1190 N. Pacific Highway
Albany, OR 97321
(catalog free)

F George W. Park Seed Company
Greenwood, SC 29647
(free catalog)

G Stokes Seeds
737 Main St.
Box 548
Buffalo, NY 14240
(free catalog)

H A World Seed Service
J.L. Hudson, Seedsman
P.O. Box 1058
Redwood City, CA 94064
($1.50 for large catalog; vegetable catalog free)

**Minor Sources**

Sometimes a particular vegetable or herb variety is available only from seed companies or nurseries *not* included in the Major Sources list. Minor sources (following) are listed as sources for vegetable and herb varieties *only* when those varieties cannot be obtained from the Major Sources.

I Burgess Seed and Plant Co.
Bloomington, IL 61701
(catalog free)

J Carroll Gardens (plants only)
P.O. Box 310
444 E. Main St.
Westminster, MD 21157
(free catalog)

K Comstock, Ferre & Co.
263 Main St.
P.O. Box 125
Wethersfield, CT 06109
(free catalog)

L Farmer Seed and Nursery Co.
Faribault, MN 55021
(free catalog)

M Henry Field Seed and
Nursery Co.
Shenandoah, IA 51602
(free catalog)

N Gardens of the Blue Ridge
(plants only)
Pineola, NC 28662
($3.00 for catalog)

O Grace's Gardens
10 Bay St.
Westport, CT 06880
($1.00 for catalog)

P Gurney Seed and Nursery Co.
Yankton, SD 57079
(free catalog)

Q Johnny's Selected Seeds,
Organic Seed and Crop
Research
Albion, ME 04910
(free catalog)

R Kitazawa Seed Co.
356 Taylor St.
San Jose, CA 95110
(free catalog)

S Orol Ledden & Sons
P.O. Box 7
Sewell, NJ 08080
(free catalog)

T Mountain Seed and Nursery
Route 1
Box 271
Moscow, ID 83843
(25¢ for catalog)

U L. L. Olds Seed Co.
P.O. Box 7790
Madison, WI 53707
(free catalog)

V Putney Nursery (plants only)
Putney, VT 05346
(free catalog)

W Redwood City Seed Co.
P.O. Box 361
Redwood City, CA 94064
(50¢ for catalog)

X Otto Richter & Sons
Box 26
Goodwood, Ontario L0C 1A0
Canada
(75¢ for catalog)

Y Southmeadow Fruit Gardens
(plants only)
2363 Tilbury Place
Birmingham, MI 48009
($8.00 for illustrated catalog;
free variety list)

Z Territorial Seed Co.
P.O. Box 27
Lorane, OR 97451
(free catalog)

AA William Tricker, Inc.
Water Garden Specialists
74 Allendale Ave.
Saddle River, NJ 07458
and 7125 Tanglewood Dr.
Independence, OH 44131
($1.00 for catalog)

BB Lilypons Water Gardens
Lilypons, MD 21717
($2.50 for catalog)

CC Tsang and Ma International
1306 Old County Rd.
Belmont, CA 94002
(free catalog)

DD Otis Twilley Seed Co.
P.O. Box 65
Trevose, PA 19047
(free catalog)

EE Van Ness Water Gardens
2460 N. Euclid Ave.
Upland, CA 91786
($2.00 for catalog)

FF Well-Sweep Herb Farm
317 Mt. Bethel Rd.
Port Murray, NJ 07865
(75¢ for catalog)

**Wholesale Sources**

Wholesale sources are listed for the benefit of researchers and seed companies who may want to obtain seed of good varieties not otherwise available, and for individual gardeners who may be able to request the seed through local garden stores.

## SOURCE LIST FOR SELECTED VEGETABLE AND HERB VARIETIES

Varieties enclosed in brackets are currently difficult or impossible for individual gardeners to order. Some of these varieties are offered only by wholesale seed companies.

GG Asgrow Seed Company
Kalamazoo, MI 49001

HH Bejo Vaden B.V.
P.O. Box 9
1722 ZG Noordscharwoude,
Holland

II Castle Seed Company
P.O. Box 877
Morgan Hill, CA 95037

JJ Elsom's Seeds Limited
Vegetable Seed Division
Spalding, PE11 1QG
England

KK Ferry Morse Seed Company
Box 100
Mountain View, CA 94040
(Seed packets sold in many stores, but no mail order)

LL Northrup King & Company
1500 Jackson St. NE
Minneapolis, MN 55413
(seed packets sold in many stores but no mail order)

MM Peto Seed Company
P.O. Box 138
Saticoy, CA 93003

NN T. Sakata & Company
C.P.O. Box Yokohama #11
Yokohama, Japan 220-91

OO Takii & Company Limited
P.O. Box 7
Kyoto Central, Kyoto
Japan 600-91

## THE VARIETIES

| | |
|---|---|
| **Amaranthus** | |
| 'Molten Fire' | C, G, H |
| 'Early Splendor' | B, C, D, F, G |
| 'Love-Lies-Bleeding, Red' | C, G, H |
| **Anise hyssop** | A, E, H |
| **Arrowhead** | |
| No variety recommendations | |
| Plants of various arrowhead species and varieties | AA, BB EE |
| **Artichoke, Globe** | |
| 'Green Globe' | B, C, F |
| **Asparagus** | |
| 'Mary Washington' | C, F, H; plants: F |
| 'Viking,' developed from 'Mary Washington' | G |
| **Basil** | |
| 'Dark Opal' | B, C, D, E, F, G, H |
| Lemon | A, F |
| Little-leaf/*O. minimum* | E, F, H |
| **Bean, Bush Snap and Wax** | |
| 'Cherokee Wax' | C, D, G |
| 'Royal Burgundy' | B, E, F, G |
| 'Royalty Purple Pod' | A |
| 'Tendercrop' | B, D, E, F |
| 'Topcrop' | B, C, F |
| **Bean, Dry** | |
| 'Montezuma Red' | E |
| **Bean, Pole** | |
| 'Kentucky Wonder' | A, B, C, D, E, F, G, H |
| 'Scarlet Runner' | A, B, C, D, G, H |
| **Beet** | |
| 'Early Red Ball' | G |
| ['Redpack' | GG] |
| ['Tendersweet Formanova' | previously I] |
| **Borage** | A, B, C, D, E, F, G, H |
| **Burnet** | A, B, E, F, H; plants: E |
| **Cabbage** | |
| 'Early Jersey Wakefield' | A, B, C, D, E, F, H |

| | |
|---|---|
| 'Ice Queen' | G |
| 'Mammoth Red Rock' | A, C, F, G, H |
| 'Red Acre' | B, C |
| 'Red Danish' | G |
| 'Red Head' | F, G |
| 'Stonehead' | B, C, E, F, G |
| 'Express' | GG |
| 'Mars No. 77' | NN |
| 'Yslanda' | HH, JJ |
| **Cabbage, Chinese** | |
| Bok Choy: most varieties, listed as: | |
| 'Chinese Pac Choi' | A, C, D, E, F, G, H |
| 'Crispy Choy' | B |
| 'Japanese White Celery Mustard' | A, E |
| 'Purple Pak Choi' | X |
| Wong Bok 'Michili' | A, B, C, F, G |
| **Calendula** | |
| 'Lemon Coronet' | K |
| 'Orange Coronet' | K |
| New varieties in 1982 catalogs may be similar to those above and are worth trial: | |
| 'Dwarf Gem' | B |
| 'Dwarf Orange' | G |
| **Carnation** | |
| 'Chabaud Giant Improved' | B, C, D, F, G, H |
| Individual 'Chabaud' colors | H |
| 'Fragrance' | C, F, G |
| 'Juliet' | B, F |
| **Carrot** | |
| 'Autumn King-Flakee' | Z |
| 'Danvers Half Long' | B, C, F |
| 'Long Imperator 58' | A, G |
| 'Royal Chantenay' | B, D, F, G |
| 'Spartan Bonus' | B, G |
| ['Scarlet Nantes Strong Top' | previously G] |
| **Celeriac** | |
| 'Alabaster' | B, X, Z |
| **Celery** | |
| 'French Dinant' | A, E |
| (possibly similar to 'Chinese Celery') | F |
| 'Summer Pascal' | C, D |
| 'Utah 52-70 R' | Q |
| 'Utah 52-70' | A, C, G |
| **Chives** | A, B, C, D, E, F, G, H |
| **Chives, Chinese/Garlic Chives** | A, B, C, E, F, H; plants: E |
| **Chrysanthemum Greens/Shungiku** | A, E, F, H |
| **Coriander** | A, B, C, E, F, H |
| **Corn** | |
| 'Bellringer' | D |
| 'Burgundy Delight' | G |
| 'Golden Cross Bantam' | B, C |
| 'Iochief' | B, C, F |
| 'Silver Sweet' | B |
| 'Stylepak' | G |
| 'Tendertreat' | DD |
| **Cucumber** | |
| 'Lemon' | A, B, C, D, E, G |
| 'Liberty' | B, E, F, G |
| 'Marketmore 70' | A, B, D, F, G |
| 'Pacer' | D |
| 'Poinsett' | C, G |
| 'Tiny Dill' | F, L |

| | |
|---|---|
| **Day Lily** | |
| *Hemerocallis flava* | plants: J |
| *H. fulva* | plants: N |
| **Dill** | |
| 'Bouquet' | A, C, D, E, F |
| Worth trial: 'Aroma' | X |
| **Eggplant** | |
| 'Early Black Egg' | Q, Z |
| 'Japanese Purple Pickling' | E |
| 'Midnight' | K |
| **Endive** | |
| 'Green Curled Ruffec'/'Green Curled'/'Ruffec' | A, B, C, D, E, F, G |
| 'Salad King' | C, F, G |
| **Escarole** | |
| 'Full Heart Batavian' | B, C, E, G, H |
| **Fennel** | |
| 'Copper' | A |
| Florence | A, B, D, E, F ('Waed-enswil'), G, H |
| Sweet | F, Q |
| **Flower Kale and Flower Cabbage/ Ornamental Kale and Ornamental Cabbage** | |
| 'Feather-leaved Coral Prince' | O |
| 'Feather-leaved Coral Queen' | O |
| 'Mixed' Flower/ Ornamental Kale | C, F, G |
| 'Mixed' Flower/ Ornamental Cabbage | C, F, G |
| 'Red on Green' Flower/ Ornamental Kale | B, D, F |
| 'White Christmas' | OO |
| 'White Lady' | OO |
| **Geranium, Scented** | |
| 'Dr. Livingston' | F |
| 'Peppermint' | F |
| 'Mixed' | F, X |
| 'Rose' | H |
| Wide selection of plants | J, FF |
| **Gourd, Fuzzy** | G, CC, O |
| Winter Melon | CC, R, O |
| **Gourd, Lagenaria** | C ('Italian Edible'), E (*L. longis-sima*) |
| **Grape** | |
| 'Concord' plants | B, I, L, U, Y |
| Selection of grape varieties (plants) | B, I, L, U, Y |
| **Husk Tomato/Husk Cherry** | C, E, G, H |
| **Kale** | |
| 'Dwarf Blue Scotch'/ 'Dwarf Blue Curled Vates' | A, B, C, D, F, H |
| 'Green Curled Scotch' | A, C, G |
| 'Westland Winter' | Z |
| ['Fribor' | HH, JJ] |
| 'Petibor' | HH |
| **Leek** | |
| 'Unique' | G |
| 'Alaska' | G |
| ['Conqueror' | previously D] |
| **Lemon Balm** | A, B, C, E, F, H; plants: B, E |

| Vegetable | Sources |
|---|---|
| **Lettuce** | |
| 'Black-seeded Simpson' | A, B, C, D, E, F, G, H |
| 'Butter King' | F, G |
| 'Grand Rapids' | B ('Burpee's Greenhart'), G |
| 'Green Ice' | B |
| 'Ithaca' | D, G |
| 'Oak Leaf' | A, B, C, D, E, F, H |
| 'Paris Island Cos' | D, G, H |
| 'Prizehead' | A, C, D ('Deep Red'), F |
| 'Royal Oak Leaf' | B |
| 'Salad Bowl' | A, B, C, D, E, F, H |
| 'Summer Bibb' | D |
| 'Valmaine Cos' | F, G |
| **Lima Bean** | |
| 'Burpee's Improved Bush' | B, C, G |
| 'Fordhook No. 242' | B, C, D, F, G, H |
| 'King of the Garden' | B, C, D, F, G, H |
| **Lovage** | A, E, F, H; plants: E |
| **Mint** | |
| Apple Mint | A; plants: E |
| Curled Mint | A, F |
| Peppermint | A, C, E, F, G, H; plants: B, E |
| Pineapple Mint | plants: E |
| Spearmint | A, C, E, F, G, H; plants: B, E |

| Vegetable | Sources |
|---|---|
| **Mizuna** | A, H |
| **Mustard Greens** | |
| 'Fordhook Fancy' | B, C |
| 'Green Wave' | C, D, G |
| 'Southern Giant Curled' | C, F, H |
| ['Prizewinner' | previously F] |
| **Nasturtium** | |
| 'Empress of India' | C, G, H |
| 'Gleam' Mixture | B, C, D, F |
| Individual 'Gleam' colors | C, F |
| **Okra** | |
| 'Red' | C, F |
| Green varieties widely available | |
| **Onion, Bunching** | |
| 'Evergreen Long White Bunching' | B, E |
| others probably equivalent to 'Evergreen Long White Bunching': | |
| 'Evergreen Bunching' | A, C |
| Evergreen White Bunching | F, H |
| 'Ishikura Long' | Q |
| 'Japanese Bunching' | D, G ('Hardy White Bunching') |
| 'Kujo Green' | Q |
| 'Common Bunching' | OO |
| **Oregano** | A, B, C, D, E, F, G; plants: B |
| Worth trial: Greek oregano | X |
| Plants of various species and varieties | J, FF |

**Parsley**

| | |
|---|---|
| 'Banquet' | D |
| 'Curlina' | F, G |
| 'Dark Green Italian' | A, D, F, G |
| 'Darki' | G |
| 'Deep Green' | S |
| 'Hamburg' | B, C, D, E, G, H |
| 'Plain'/'Italian' | A, B, E, H |

**Parsnip**

| | |
|---|---|
| 'Harris Early Model' | C |
| 'Harris Model' | D, G |
| 'Hollow Crown' | A, B, C, F, H |
| 'Improved Hollow Crown' | G |

**Pea**

No variety recommendation

**Pepper**

| | |
|---|---|
| 'Canape' | D, E, G |
| 'Cherry Sweet' | B, C, D, E, F, G, H |
| 'Fiery Festival' | D |
| 'Fushimi Long Green' | W |
| 'Golden Calwonder' | B, G, H |
| 'Hot Hungarian Yellow Wax' | A, B, D, E, F, G, H |
| 'Italian Sweet' | I, Z |
| 'Keystone Resistant Giant' | C, G, Z |
| 'Long Thin Cayenne' | C, F ('Cayenne Long Red Slim') |
| 'New Ace' | B |
| 'Red Chile' | Z |
| 'Shepherd' | G ('Super Shepherd') |
| 'Sweet Banana' | B, C, F, G |
| 'Early Bountiful' | NN |
| ['Petite Yellow Sweet' | previously MM] |

**Poppy**

| | |
|---|---|
| 'Shirley Mixed Single' | B, G |

**Radish, Fall**

| | |
|---|---|
| 'All Seasons White' | B, C |
| 'Sakurajima Giant' | E, H, W, probably C ('White Giant') |

**Rhubarb**

No variety recommendation

**Rose**

No variety recommendation

| | |
|---|---|
| Seed: | |
| *Rosa canina* | H |
| *R. eglanteria* | H |
| *R. rugosa alba* | H |
| *R. rugosa* single pink | H |
| *R. spinosissima* | H |
| Plants: | |
| *Rosa pomifera* | L |
| *R. rugosa:* 5 varieties | L |

| | |
|---|---|
| **Rosemary (erect)** | A, C, E, F, G, H; plants: E |
| Creeping | X; plants: E |
| Wide selection of plants of different varieties | J, FF |
| **Sage** | A, B, E, F, G, H; plants: B, E |
| 'Broadleaf' | C, D |
| Plants of many species and varieties | J, FF |
| **Savory, Summer** | A, B, C, D, E, F, G, H |
| **Savory, Winter** | A, E, F, H |
| Plants of different varieties | E, FF |

| | |
|---|---|
| **Sesame** | C, E, F, H |
| **Soybean** | |
| 'Envy' | A, Q, Z |
| 'Fiskeby V' | A, G, Q |
| 'Pickett' | F |
| 'Prize' | B, E |
| [Frostbeater | previously B] |
| **Spinach, New Zealand** | A, B, C, D, E, F, G, H |
| **Squash, Summer** | |
| 'Aristocrat' | B, E, F |
| 'Blackini' | P |
| 'Burpee Hybrid Zucchini' | B |
| 'Eldorado' | D |
| 'Gold Rush,' which resembles 'Eldorado' | B, E, F, G |
| 'Scaloppini' | B, E, F, G, P |
| 'Castle Black' | II |
| 'Castle Verde' | II |
| ['Diplomat' | MM] |
| ['Golden Eagle' | previously LL] |
| **Squash, Winter, and Pumpkin** | |
| 'Delicata' | A, G |
| 'Golden Delicious' | D, G |
| 'Gold Nugget' | A, C, D, F, G |
| 'Huicha' | H |
| 'Royal Acorn' | B, E, G, H |
| 'Triple Treat' pumpkin | B |
| 'Vegetable Gourd' | C, F, P |
| **Sunflower** | |
| 'Mammoth'/ 'Mammoth Russian' | A, B, C, G |
| 'Sundak' | A, Q |
| Worth trial: 'Sunbird' | B |
| **Swiss Chard** | |
| 'Rhubarb' | A, B, C, D, E ('Ruby'), F, G, H |
| **Thyme** | |
| Creeping/*T. serpyllum* | A, C, E, F, H |
| Winter Thyme 'Dwarf Compact' | F |
| Plants of various species and varieties | E, J, FF |
| **Tomato** | |
| 'Dwarf Champion' | G |
| 'Floramerica' | B, D, E, F, G |
| 'Gardener's Delight' | B, Q |
| 'Heinz 1370' | G, DD |
| 'Kootenai' | T |
| 'Red Cherry' | A, B, C, G, H |
| 'Red Cherry Small' | D |
| 'Redpak' | S |
| 'Roma' | B, C, D, E, F, G, H |
| 'San Marzano' | B, E, G, H |
| 'Small Fry' | B, D, F, G |
| 'Yellow Plum' | B, C, E, F, G, H |
| ['Saladmaster' | previously T] |
| **Violet, Sweet** | |
| No variety recommendation | F, H |
| Plants of different varieties | B, H, J |
| Plants of native violet species | V |
| **Water Chestnut** | Tubers from EE, oriental grocery stores |

## SOURCE LIST FOR SELECTED FRUITS AND NUTS

Addresses not given here are listed earlier in this Appendix.

**Plants**

Bountiful Ridge Nurseries
Princess Anne, MD 21853
(free catalog)

Burgess Seed and Plant Company

Burpee Seed Company

Farmer Seed Company

Henry Field Seed Company

Louis Gerardi
R. R. 1, Box 146
O'Fallon, IL 62269
(many nuts)

Southmeadow Fruit Gardens

Stark Brothers
Louisiana, MO 63353
(free catalog)

**Seeds**

Abundant Life Seed Foundation (northwest natives)

Redwood City Seed Company

A World Seed Service

## SOURCE LIST OF SOME "POSSIBILITIES TO EXPLORE'
(From the list at the end of chapter 3)

Addresses not given here are listed earlier in this Appendix.

Carroll Gardens (herbs and perennial flowers)

Gardens of the Blue Ridge (native plants)

Three Springs Fisheries (water plants)

William Tricker (water plants)

Van Ness Water Gardens

Well-Sweep Herb Farm

**Seeds**

Abundant Life Seed Foundation (herbs and natives)



Redwood City Seed Company (rare edibles)

Otto Richter & Sons (herbs)

A World Seed Service (encyclopedic!)

## ORNAMENTAL-EFFECTIVENESS EVALUATION

Information from Vegetable and Herb Trial Gardens 1977 and 1978, Oregon State University Vegetable Research Farm, Corvallis, Oregon.

Some varieties were grown for one year, some for two years. Numbers listed, from − 3 to 10, reflect overall ornamental effectiveness of plants; qualities of crop yield and flavor are not included here.

**Amaranthus**
'Love-Lies-Bleeding,' 10; 'Molten Fire,' 10.

**Anise hyssop,** 10.

**Arrowhead** (no definite variety information), 9.

**Artichoke**
'Green Globe,' 10.

**Asparagus** (no definite variety information), 10.

**Basil**
'Dark Opal,' 9; 'Lettuce Leaf,' 1; 'Little Leaf,' 10; 'Sweet,' 7.

**Bean, Bush Snap**
'Contender,' 4; 'Early Gallatin,' 7; 'Green Crop,' 6; 'Provider,' 5; 'Royal Burgundy,' 8; 'Royalty Purple Pod,' 7; 'Sprite,' 4; 'Tendercrop,' 7; 'Topcrop,' 7.

**Bean, Bush Wax**
'Buerre de Rocquencourt,' 6; 'Cherokee Wax,' 6.

**Bean, Dry**
'California Red Kidney,' 5; 'Montezuma Red,' 6; 'Pinto,' 3; 'Red Peanut,' −2; 'Vermont Cranberry,' 3.

**Bean, Pole**
'Kentucky Wonder,' 7; 'Purple Podded,' 1; 'Romano'/'Italian Pole,' 5.

**Beet**
'Beets for Greens,' 3; 'Burgundy King,' 6; 'Burpee Golden,' −2; 'Burpee's Red Ball,' 0; 'Crosby Egyptian,' 7; 'Cylinder Long Red,' 5; 'Detroit Dark Red,' 5; 'Early Blood Turnip,' 4; 'Early Bunching,' 5; 'Early Red Ball,' 8; 'Early Staysgreen,' 4; 'Early Wonder,' 4; 'Fire Chief,' 5; 'Formanova,' 5; 'Globe,' 7; 'Green Top Bunching,' 4; 'Honey Red,' 4; 'Little Ball,' 5; 'Little Egypt,' 4; 'Lutz Green Leaf,' 3; 'New Globe,' 5; 'Red Ball,' 1; 'Red Cross,' 0; 'Redpack,' 8; 'Ruby Queen,' 6; 'Spinel,' 4; 'Spring Red,' 4; 'Stokes Special Early,' 3; 'Tendersweet Formanova,' 8; 'Vermilion,' 4; 'Wonder,' 3; 'Yellow Eckendorf,' 2.

**Borage,** 8.

**Burnet,** 8.

**Cabbage, Green**
'Copenhagen Market,' 5; 'Defender,' 4; 'Early Jersey Wakefield,' 8;'Enterprise,' 7; 'Express,' 8; 'Golden Acre,' 5; 'Hi Dri No. 6428,' 8; 'Market Victor,' 5; 'Mars No. 77,' 10; 'Portugal,' 3; 'Rapid Ball,' 1; 'Stonehead,' 8; 'Sun-up,' 3; 'Tastie Hybrid,' 3.

**Cabbage, Red**
'Langendijk Extra Early,' 6; 'Large Red Dutch Storage,' 4; 'Mammoth Red Rock,' 10; 'Red Acre,' 7; 'Red Danish,' 9; 'Red Debut,' 7; 'Red Head,' 9; 'Red Meteor,' 5; 'Red Storage,' 7; 'Red Storage No. 4004,' 1; 'Ruby Ball,' 6.

**Cabbage, Savoy**
'Ice Queen,' 10; 'Savoy Ace,' 7; 'Savoy Chieftain,' 7; 'Savoy King,' 8; 'Yslanda,' 10+.

**Cabbage, Chinese: Bok Choy**
'Crispy Choy Loose-leaved,' 3; 'Pak Choi Chinese,' 6; 'Pak Choi Japanese Giant White Celery Mustard,' 4; 'Pak Choi Japanese White Celery Mustard,' 4.

**Cabbage, Chinese: Wong Bok**
'Early Hybrid G,' 1; 'Michili,' 5; 'Nagaoka No. 2,' 0; 'Oriental King No. 18,' 1; 'Round-leaved Santo,' 4; 'Serrated Leaf Santo,' 4; 'Spring A-1,' −2; 'Tip-Top No. 12,' 1; 'Tropical Delight No. 23,' 1; 'WR 65 Days,' 4.

**Calendula**
*C. officinalis,* 7; 'Orange Coronet,' 8; 'Orange King,' 7; 'Pacific Beauty Mixture,' 6.

**Carnation**
'Chabaud Giant Improved,' 10; 'Fragrance,' 9; 'Juliet,' 10.

**Carrot**
'Autumn King,' 10+; 'Bunny Bite,' 3; 'Candypack,' 8; 'Chantenay Red Cored,' 8; 'Danvers Half Long,' 10; 'Dominator,' 10; 'Gold King,' 6; 'Gold Pak,' 7; 'Grenadier,' 5; 'Kinko Cross 6,' 4; 'Kinko Cross 8,' 4; 'Lance,' 10; 'Long Imperator, No. 58,' 9; 'Mini Express,' 9; 'Parisian Rondo,' 2; 'Planet,' 3; 'Royal Chantenay,' 9; 'Scarlet Nantes Strong Top,' 9; 'Short 'n' Sweet,' 4; 'Spartan Bonus,' 10; 'Target,' 7.

**Celeriac**
'Alabaster,' 9; 'Marble Ball,' 6.

**Celery**
'Burpee's Fordhook,' 8; 'French Dinant,' 10+; 'Summer Pascal,' 9½; 'Tendercrisp,' 8; 'Utah 52-70 R,' 10.

**Chives,** 8.

**Chives, Chinese/Garlic Chives**
'Broadleaf,' 9; 'Flowering Leek Tenderpole,' 8.

**Chrysanthemum Greens/Shungiku,** 7.

**Coriander,** 9.

**Corn**
'Bellringer,' 8; 'Burgundy Delight,' 8; 'Candystick,' 7; 'Early Sunglow Hybrid Yellow,' 4; 'Fanfare,' 7; 'Golden Beauty,' 6; 'Golden Cross Bantam,' 8; 'Golden Midget,' 3; 'Iochief,' 9; 'Jubilee,' 7; 'Midnight Snack,' 1; 'Queen Anne,' 0; 'Quicksilver,' 7; 'Rapidpak,' 7; 'Reliance,' 7; 'Ruby Gem,' 7; 'Seneca Chief,' 7; 'Silver Sweet,' 6;

'Spring Gold,' 6; 'Stylepak,' 10; 'Sugar Loaf,' 6; 'Sweet Suc,' 5; 'Tendertreat,' 10+; 'Tokay Sugar,' 6.

Other varieties tested: 'Buttercorn,' 'Candy Man,' 'Carmelet,' 'Earliking,' 'Early Sunglow,' 'Honeycomb,' 'Pageant,' 'Patriot,' 'Snopak,' 'Starlet,' 'White Crisp.'

**Cucumber**
'Burpless Hybrid,' 1; 'Bush Whopper,' 8; 'Calypso,' 5; 'Cherokee 7,' 7; 'County Fair,' 3; 'Earlyset No. 36,' 1; 'Green Star,' 5; 'Improved Pioneer,' 3; 'Jet Set No. 20,' −2; 'Kyoto 3 Feet,' 4; 'Lemon,' 9; 'Liberty Hybrid,' 7; 'Marketmore No. 70,' 8; 'Pacer,' 10; 'Park's Whopper,' 9; 'Patio Pik,' 6; 'Peppi,' 1; 'Philly,' 7; 'Pickledilly,' 7; 'Poinsett,' 7; 'Pot Luck,' 7; 'Saladin,' 7; 'Salty No. 335,' 7; 'Shamrock,' 6; 'Slicerite,' 4; 'Southern Cross,' 1; 'Spacemaster,' 7; 'Sprint,' 6; 'Straight Eight,' 8; 'Superslice,' 7; 'Tamu Triple Cross,' 6; 'Tiny Dill,' 9; 'Triple Mech,' 7; 'Trispear,' 4; 'Triumph,' 7; 'Victory,' 8; 'West Indian Gherkin,' 5.

**Day Lily** (no definite variety information), 9.

**Dill**
'Bouquet,' 8; 'Long Island Mammoth,' 3.

**Eggplant**
'Albino,' 4; 'Apple Green,' 1; 'Black Beauty,' 7; 'Blackjack,' 3; 'Blacknight,' 7; 'Burpee's Hybrid,' 3; 'Dusky,' 6; 'Early Black Egg,' 8; 'Golden Dwarf,' 5; 'Ichiban,' 5; 'Jersey King,' 7; 'Large-fruited No. 30,' 4; 'Midnite,' 9; 'Mission Bell,' 5; 'Money Maker,' 6; 'Morden Midget,' 2; 'Nagaoka Long Black,' 3; 'Japanese Purple Pickling,' 8; 'Royal Knight,' 8; 'Slim Jim,' 4; 'Small-fruited Long Tom No. 4,' 7; 'Tenderette,' −2; 'White Beauty,' 4.

**Endive**
'Green Curled Ruffec,' 9; 'Salad King,' 9.

**Escarole**
'Batavian Full Heart,' 7.

**Fennel**
'Copper,' 10+; 'Florence (Finocchio),' 9; 'Perfection,' 8; 'Sweet,' 9; *vulgare,* 8.

**Flower Kale/Cabbage**
'Feather-leaved Coral Prince,' 10; 'Feather-leaved Coral Queen,' 10; 'Mixed (Stokes),' 9;'Osaka Round-leaved Sekito,' 10; 'Red on Green' (Harris),' 10; 'Rose Bouquet,' 9; 'Sekito,' 5; 'Tokyo Round-leaved White,' 7; 'White Christmas,' 10; 'White Lady,' 10; 'White on Green' (Harris), 9.

**Geranium, Scented**
'Peppermint,' 9; 'Rose,' 9.

**Gourd, Lagenaria**
'Calabash Indian,' 9; 'Calabash Large Round,' 7; 'Cucuzzi,' 7; 'Giant Bottle,' 8; 'Lagenaria Serpent,' 9.

**Gourd, Fuzzy**
'Long Shape,' 10.

**Grape Leaves**
'Concord,' 10+.

**Husk Tomato/Husk Cherry,** 7.

**Kale**
'Arpad F1,' 10; 'Dwarf Blue Curled Vates,' 10; 'Fribor F1,' 10; 'Green Curled Scotch,' 9; 'Kailann White-flowered,' 6; 'Konserva,' 7; 'Pentland Brigg,' 5; 'Petibor F1,' 10+; 'Siberian,' 6; 'Tall Green Curled,' 6; 'Westlandse Winter,' 9.

**Leek**
'Alaska,' 9; 'Artico,' 9; 'Broad London/American Flag,' 8; 'Conqueror,' 10+; 'Unique,' 10.

**Lemon Balm,** 7.

**Lettuce**
'Arctic King,' 1; 'Bibb,' 3; 'Black-seeded Simpson,' 9; 'Burpee's Greenhart Grand Rapids,' 9; 'Buttercrunch,' 2; 'Butter King,' 9; 'Continuity,' 10+; 'Dark Green Boston,' 5; 'Dark Green Cos,' 9; 'Deer Tongue,' 7; 'Fortune,' 8; 'Grand Rapids,' 9; 'Great Lakes No. 659,' 7; 'Green Ice,' 10; 'Green Mignonette,' 3; 'Hot Weather,' 10; 'Ithaca,' 7; 'Little Gem Cos,' 4; 'Lobjoit's Green Cos,' 7; 'Mesa No. 659,' 4; 'Mignonette Bronze,' 7; 'Oak Leaf,' 7; 'Paris Island Cos,' 10; 'Paris White Cos/ Trianon,' 7; 'Pennlake,' 4; 'Prizehead,' 10; 'Red Salad Bowl,' 10; 'Royal Oak Leaf,' 8; 'Ruby,' 7; 'Salad Bowl,' 9; 'Salad Trim,' 9; 'Sigmadeep,' 8; 'Summer Bibb,' 8; 'Suzan,' 4; 'Tendercrisp,' 8; 'Tom Thumb,' 1; 'Winter Density,' 7.

**Lima Bean**
'Burpee's Improved,' 8; 'Fordhook No. 242,' 8; 'Henderson's Bush,' 7; 'Improved Kingston,' 8; 'Jackson Wonder,' 7; 'King of the Garden,' 9.

**Lovage,** 8.

**Mint**
'Curled Mint,' 7; 'Peppermint,' 7; 'Spearmint,' 8.

**Mizuna,** 10.

**Mustard Greens**
'Fordhook Fancy,' 8; 'Florida Broadleaf,' 6; 'Green Wave,' 9; 'Mustard Spinach Late Shirona,' 6; 'Prizewinner Curled Long,' 10; 'Southern Giant Curled,' 8; 'Tendergreen India Mustard,' 5.

**Nasturtium**
'Dwarf Dark-leaved Mixed,' 10; 'Empress of India,' 10+; 'Gleam Mixture,' 10; 'Jewel Mixture,' 9; 'Single Tall Climbing Mixed,' 9; 'Whirlybird Mixture,' 8; 'Whirlybird Scarlet,' 9.

**Okra** (while healthy)
'Clemson Spineless,' 6; 'Dwarf Long Pod Green,' 6; 'Emerald,' 7; 'Red,' 9.

**Onion, Bunching**
'Beltsville Bunching,' 4; 'Common Bunching,' 10; 'Evergreen Long White Bunching,' 10; 'Ishikura Long Bunching,' 10+; 'Japanese Bunching,' 10; 'Kujo Green,' 9; 'Perfecto Blanco,' 3; 'White Bunching/White Lisbon,' 3.



**Oregano,** 9.

**Parsley**
'Banquet,' 10; 'Bravour,' 7; 'Champion Moss Curled,' 7; 'Curlina,' 10; 'Dark Green Italian,' 10+; 'Darki,' 9; 'Deep Green,' 10; 'Delicata Original,' 7; 'Evergreen Double Curled,' 8; 'Extra Curled Dwarf,' 9; 'Extra Triple Curled,' 6; 'Hamburg Thick Root,' 9; 'Krousa,' 9; 'Moskrul Triplex,' 8; 'Paramount,' 9; 'Plain Hardy Italian,' 10.

**Parsnip**
'Harris Early Model,' 9; 'Improved Hollow Crown,' 9.

**Pea** (no definite variety information), 7.

**Pepper**
'Anaheim M,' 4; 'Atlas,' 3; 'Bell Boy,' 4; 'Bigpak,' 4; 'Blanco del Pais,' 3; 'Cadice,' 5; 'California Wonder 300,' 6; 'Caloro,' 6; 'Canape,' 8; 'Cayenne,' 5; 'Cayenne Large Red Thick,' 6; 'Cherry Sweet,' 6; 'Cheesepak,' 4;'Cubanelle,' 4; 'Dutch Treat,' 2; 'Earliest Red Sweet,' 4; 'Early Bountiful F1' 8; 'Early Canada Bell,' 5; 'Early Set,' 4; 'Early Sweet Banana,' 8; 'Fiery Festival,' 10; 'Fushimi Long Green,' 8; 'Golden Bell,' 6; 'Golden Calwonder,' 8; 'Goldspike,' 8; 'Hot Cayenne Thick,' 6; 'Hot Hungarian Yellow Wax,' 8; 'Italian Sweet,' 8; 'Keystone Resistant Giant,' 9; 'Lady Bell,' 4; 'Lamuyo,' 5; 'Long Thin Cayenne,' 7; 'Long Yellow Sweet,' 4; 'Mercury,' 6; 'Midway,' 4; 'Morgold,' 5; 'New Ace,' 7; 'New Mexico Big Jim,' 3; 'Nosegay,' 5; 'Parkwonder,' 4; 'Petite Yellow Sweet,' 8; 'Pimento L,' 5; 'Pimento Select,' 3; 'Pinocchio,' 1; 'Pip,' 6; 'RCEC,' 4; 'Red Cherry Large Hot,' 4; 'Red Cherry Small Hot,' 4; 'Red Chile' (Germaine), 9; 'Red Chile' (Peto), 10+; 'Roumanian Sweet,' 3; 'Shepherd,' 8; 'Sweet Banana,' 7; 'Sweet Chocolate,' 3; 'Vinedale,' 6; 'Whopper,' 7; 'Yellow Belle,' 4; 'Yellow Castle,' 5; 'Yellow Cheese Pimento,' 6; 'Yolo Wonder L,' 5.

**Poppy**
'Shirley Mixed Single,' 9.

**Radish, Fall**
'All Seasons White,' 8; 'China Rose,' 3; 'Round Black Spanish,' 3; 'Sakurajima' (Burgess), 6; 'Sakurajima Mammoth' (Sakata), 9; 'Summer Cross Hybrid,' 5; 'White Chinese/Celestial,' 5; 'Winter King Miyako,' 6.

**Rhubarb** (no definite variety information), 8.

**Rose** (no definite variety information), 9.

**Rosemary** (erect), 10.

**Sage**
'Broadleaf,' 9.

**Savory, Summer,** 10.

**Savory, Winter,** 8.

**Sesame** (when healthy), 10.

**Soybean**
'Akita Early,' 4; 'Altona,' 3; 'Early Green,' 5; 'Envy,' 7; 'Fiskeby V,' 6; 'Frostbeater,' 7; 'Hakucho Extra Early,' 3; 'Improved Hakucho,' 4; 'Mikawashima Green,' −2; 'Okuhara Early,' 4; 'Pickett,' 8; 'Prize,' 6; 'Sodefuri Green,' 5; 'Takii's Extra Early,' 7.

**Spinach**
'New Zealand,' 7.

**Squash, Summer**
'Ambassador,' 7; 'Baby Crookneck,' −2; 'Black Eagle,' 5; 'Blackini,' 9; 'Blackjack,' 5; 'Burpee Hybrid Zucchini,' 8; 'Butterpak,' 1; 'Castleblack,' 10; 'Castlegold,' 7; 'Castle Verde,' 10; 'Chefini,' 4; 'Cocozelle (Long Type),' 9; 'Creamy,' 1; 'Daytona,' 1; 'Diamont,' 7; 'Diplomat,' 8; 'Early White Bush Scallop,' 4; 'Eldorado,' 8; 'Goldbar,' 4; 'Golden Eagle,' 7; 'Goldneck,' 3; 'Goldstrike,' 4; 'Goldzini,' 4; 'Greyzini,' 4; 'Patty Green Tint,' 1; 'President,' 1; 'Saint Pat Scallop,' 3; 'Scaloppini Hybrid,' 6; 'Seneca Prolific,' 2; 'Summer Sun,' 1; 'Sundance,' 1; 'Tatume,' 5; 'Vegetable Marrow Bush,' 4.

**Squash, Winter, and Pumpkin**
'Baby Butternut,' 4; 'Banana,' 4; 'Blue Kuri,' 3; 'Bush Butternut,' 1; 'Bush Ebony Acorn,' 5; 'Bush Table Queen,' 6; 'Buttercup,' 4; 'Butternut,' 4; *C. ficifolia,* 9; 'Delica,' 3; 'Delicata,' 10; 'Early Butternut,' 3; 'Eat All,' 7; 'Ebony,' 7; 'Emerald,' 4; 'Funny Face,' 2; 'Golden Cushaw,' 5; 'Golden Delicious,' 8; 'Gold Nugget,' 9; 'Hiyuga Early Black,' 7; 'Huicha,' 10; 'Improved Hubbard,' 4; 'Jack o' Lantern,' 4; 'Kikuza Early White,' 6; 'Kindred,' 7; 'Lady Godiva,' 3; 'Melon Squash,' 7; 'Mixta Gold,' 6; 'New England Blue Hubbard,' 4; 'Red Kuri,' 5; 'Royal Acorn,' 7; 'Small Sugar,' 6; 'Spirit,' 4; 'Sweet Dumpling,' 6; 'Sweet Mama,' 3; 'Sweetnut,' 4; 'Table King,' 6; 'Table Queen,' 5; 'Tetsukabuto,' 9; 'Triple Treat,' 8; 'Vegetable Gourd,' 8; 'Vegetable Spaghetti,' 3; 'Waltham Butternut,' 3; 'Young's Beauty,' 4.

**Sunflower**
'Mammoth,' 9; 'Sundak,' 9.

**Swiss Chard**
'Common Green,' −2; 'Fordhook Giant,' 3; 'Lucullus,' 4; 'Perpetual,' 3; 'Rhubarb,' 9; 'White King,' 5.

**Thyme**
'Creeping,' 10+; 'Winter,' 9; 'Winter Dwarf Compact,' 10.

**Tomato**
'Basketpak,' 6; 'Beefmaster,' 1; 'Better Boy,' −2; 'Bigset,' 4; 'Bitsy,' −2; 'Bonus,' 5; 'Burgess Stuffing,' 5; 'Burpee's Pixie,' 0; 'Bush Beefsteak,' 1; 'Dwarf Champion,' 6; 'Early Cherry,' −2; 'Early Girl,' 3; 'Early Subarctic,' −3; 'Floramerica,' 6; 'Gardener's Delight,' 6; 'Golden Delight,' 4; 'Heinz 1350,' 4; 'Heinz 1370,' 7; 'Heinz 1439,' 7; 'Hybrid Early Salad,' 1; 'IPB,' −3; 'Jetfire,' 0; 'Jubilee,' 4; 'Kootenai,' 7; 'Moira,' 1; 'Monte Carlo,' 6; 'New Yorker,' −2; 'Orange Queen,' 3; 'Patio,' 2; 'Ponderosa Pink,' −2; 'Presto,' 0; 'Red Cherry,' 7; 'Red Cherry Large,' 4; 'Red Cherry Small,' 7; 'Redpak,' 6; 'Red Pear,' 5; 'Roma,' 6; 'Royal Chico,' 5; 'Royal Flush,' 4; 'Saladette,' 4; 'Saladmaster,' 8; 'San Marzano,' 6; 'Sigmabush,' 1; 'Small Fry,' 6; 'Speedy,' 4; 'Spring Giant,' 4; 'Springset,' 0; 'Starshot,' 1; 'Sugar Lump,' 1; Sunray,' 5; 'Swift,' 1; 'Tiny Tim,' −2; 'Toy Boy,' 4; 'Tumblin' Tom,' 4; 'Willamette,' 4; 'Yellow Pear,' 1; 'Yellow Plum,' 6.



**Violet** (no definite variety information), 8.

**Water Chestnut,** 8.

# Selected Bibliography

Books marked * are especially suitable as supplements to the material in this book.

## Background Reading

Anderson, Edgar. *Plants, Man and Life.* Berkeley and Los Angeles: University of California Press, 1969.

Berrall, Julia S. *The Garden: An Illustrated History.* New York: Viking Press, 1966.

The Findhorn Community. *The Findhorn Garden.* New York: Harper & Row, 1975.

Fukuoka, Masanobu. *The One Straw Revolution.* Emmaus, Pa.: Rodale Press, 1978.

Hatton, Richard G. *Handbook of Plant and Floral Ornament: From Early Herbals,* 1909. Reprint. New York: Dover Publications, 1960.

Lappé, Frances Moore. *Diet for a Small Planet.* New York: Friends of the Earth/Ballantine Books, 1971.

McHarg, Ian. *Design with Nature.* Garden City, N.Y.: Doubleday/ Natural History Press, for the American Museum of Natural History, 1969.

## Landscape Design References

Brookes, John. *Room Outside: A New Approach to Garden Design,* 2nd ed. New York: Penguin, 1979.

*Creasy, Rosalind. *The Complete Book of Edible Landscaping.* San Francisco: Sierra Club Books, 1982.

*Eckbo, Garrett. *Art of Home Landscaping.* New York: McGraw-Hill Book Company, 1956.

*———. *Home Landscape: The Art of Home Landscaping,* rev. enl. ed. New York: McGraw-Hill Book Company, 1978.

*Grant, John A., and Grant, Carol L. *Garden Design Illustrated.* Seattle: University of Washington Press, 1954.

Koberg, Don, and Bagnall, Jim. *The Universal Traveller.* Los Altos, California: William Kaufmann, 1976.

Robinette, Gary. *Landscape Planning for Energy Conservation.* Reston, Virginia: Environmental Design Press, 1977.

———. *Plants/People/ and Environmental Quality.* Washington, D.C.: U.S. Department of the Interior, National Park Service, and American Society of Landscape Architects Foundation, 1972.

Rose, James C. *Creative Gardens.* New York: Reinhold Publishing Corporation, 1958.

Simonds, John Ormsbee. *Landscape Architecture: The Shaping of Man's Natural Environment.* New York: McGraw-Hill Book Company, 1961.

## Plants and Gardening References

Baggett, James R. Class materials on vegetables for "Systematics of Fruits and Vegetables" course. Mimeographed. Corvallis, Ore.: Oregon State University, 1977.

Bailey, Liberty Hyde, and Bailey, Ethel Zoë. *Hortus Third.* Revised and expanded by staff of L. H. Bailey Hortorium of Cornell University. New York: Macmillan Publishing Company, 1976.

Berg, Gordon, ed. *Farm Chemicals Handbook.* Willoughby, O.: Meister Publishing Company, 1980. Published annually.

Bianchini, Francesco, and Corbetta, Francesco; illustrated by Marilena Pistoia. *The Complete Book of Fruits and Vegetables.* New York: Crown Publishers, 1973.

Boston Urban Gardeners. *Lead Booklet.* National Center for Appropriate Technology, Publications Department TMEN, P.O. Box 3838, Butte, Montana 59702 (50¢).



Bryan, John E., and Castle, Coralie. *The Edible Ornamental Garden.* San Francisco: 101 Productions, 1974.

*Co-Evolution Quarterly.* P.O. Box 428, Sausalito, California 94965.

Colebrook, Binda. *Winter Gardening in the Maritime Northwest.* Arlington, Washington: Tilth, 1977.

Crockett, James Underwood; Tanner, Ogden; and Editors of Time-Life Books. *Herbs.* Alexandria, Virginia: Time-Life, 1977.

DeCrosta, Tony. "How Heavy Metals Pollute Our Soils." *Organic Gardening,* June, 1981.

Dennis, La Rea J. *Name Your Poison: A Guide to Cultivated and Native Oregon Plants Toxic to Humans.* Corvallis, Ore.: Oregon State University Bookstore, 1972.

Doty, Walter L., ed. *All About Vegetables.* San Francisco: Ortho Division—Garden & Home, Chevron Chemical Company, 1973.

DUMP HEAP, *The Journal of Diverse Unsung Miracle Plants for Healthy Evolution Among People.* Quarterly. 371 Irwin St., San Rafael, California 94901.

Farralones Institute. *The Integral Urban House.* San Francisco: Sierra Club Books, 1979.

Fernald, Merritt Lyndon, and Kinsey, Alfred Charles. *Edible Wild Plants of Eastern North America.* Revised by Reed C. Collins. New York: Harper & Row, 1958.

*Foster, Gertrude B. *Herbs for Every Garden.* New York: E. P. Dutton & Company, 1966.

Fowler, Cary. *Graham Center Seed Directory.* Wadesboro, North Carolina: Frank Porter Graham Center, 1979.

Furlong, Marjorie, and Pill, Virginia. *Wild Edible Fruits and Berries.* Healdsburg, California: Naturegraph Publishers, 1974.

*Gardens for All News.* 180 Flynn Ave., Burlington, Vermont 05401.

Gerard, John. *Leaves from Gerard's Herball.* Arranged by Marcus Woodward, 1931. (Original, 1597). Reprint. New York: Dover Publications, 1969.

Gessert, Kate Rogers. "Encyclopedia of Tropical and Subtropical Fruits." Master's thesis, Bank Street College of Education, New York, 1976.

———. "Mixing Food Plants with Ornamentals." *American Horticulturist,* Early Spring 1976.

Gibbons, Euell. *Stalking the Healthful Herbs.* Field Guide ed. New York: David McKay Company, 1970.

Graf, Alfred Byrd. *Exotic Plant Manual,* 2d ed. East Rutherford, New Jersey: Roehrs Company, 1970.

Grant, John A., and Grant, Carol A. *Trees and Shrubs for Pacific Northwest Gardens.* Palo Alto, California: Pacific Books, 1943.

Grieve, Mrs. M. *A Modern Herbal.* 2 vols., 1931. Reprint. New York: Dover Publications, 1971.

Harrington, Geri. *Grow Your Own Chinese Vegetables.* New York: Macmillan Publishing Company, Collier Books, 1978.

Harris, Ben. *Make Use of Your Garden Plants.* Barre, Massachusetts: Barre Publishing Company, 1978.

Hay, Roy, and Synge, Patrick M. *The Color Dictionary of Flowers and Plants for Home and Garden.* New York: Crown Publishers, in collaboration with the Royal Horticultural Society, 1969.

Hedrick, Ulysses Prentiss. *Cyclopedia of Hardy Fruits.* New York: Macmillan Company, 1938.

Johnson, Hugh. *The International Book of Trees.* New York: Simon & Schuster, 1973.

Kraft, Ken and Kraft, Pat. *Exotic Vegetables: How to Grow and Cook Them.* New York: William Morrow & Company, 1977.

———. *Fruits for the Home Garden.* New York: William Morrow & Company, 1968.

Lent, Joseph M.; McGourty, Frederick, Jr.; and Dietz, Marjorie J., eds. *The Home Vegetable Garden: A Handbook.* Brooklyn Botanic Garden Handbook No. 69. Brooklyn, New York: Brooklyn Botanic Garden, 1972.

Logsdon, Gene. *Small-scale Grain Raising.* Emmaus, Pa.: Rodale Press, 1977.

McGourty, Frederick, Jr., ed. *Perennials and Their Uses.* Brooklyn Botanic Garden Record; Plants and Gardens, Autumn 1978.

Martin, W. Keble. *The Concise British Flora in Color.* London: Sphere Books, with Ebury Press and Michael Joseph, 1965.

Masefield, G. B.; Wallis, M.; Harrison, S. G.; and Nicholson, B. E. *The Oxford Book of Food Plants.* London: Oxford University Press, 1969.

Masters, Charles O. *Encyclopedia of the Water Lily.* Neptune City, New Jersey: T. F. H. Publications, 1974.

Mollison, Bill. *Permaculture Two: Practical Design for Town and Country in Permanent Agriculture.* Tasmania, Australia: Tagari Books, 1979.

*——— and Holmgren, David. *Permaculture One: A Perennial Agriculture for Human Settlements.* Ealing, England: Corgi Books and Transworld Publishers, 1977.

*Morton, Julia F. *Herbs and Spices: A Golden Guide.* New York: Golden Press, 1976.

Muenscher, Walter Conrad. *Poisonous Plants of the United States,* rev. ed. New York: Macmillan Publishing Company, Collier Books, 1975.

Nehrling, Arno, and Nehrling, Irene. *The Picture Book of Annuals.* New York: Hearthside Press, 1966.

———. *The Picture Book of Perennials.* New York: Hearthside Press, 1964.

New York Unit of the American Herb Society, editorial committee; Van Brunt, Elizabeth R., chairman; Nelson, Peter K.; and McGourty, Frederick, Jr., eds. *Handbook on Herbs.* Brooklyn Botanic Garden Handbook No. 27. Brooklyn, New York, 1972.

Nitschke, Robert A. *Southmeadow Fruit Gardens: Choice and Unusual Fruit Varieties for the Connoisseur and Home Gardener,* 1976 ed. Birmingham, Michigan: Southmeadow Fruit Gardens, 1976.

Olkowski, Helga, and Olkowski, William. *The City People's Book of Raising Food.* Emmaus, Pennsylvania: Rodale Press, 1975.

*Organic Gardening and Farming Magazine.* 33 E. Minor St., Emmaus, Pennsylvania.

Ortho Books, Editorial Staff. *All About Growing Fruits and Berries,* ed. Will Kirkman. San Francisco: Ortho Books, Chevron Chemical Company, 1976.

———. *All About Tomatoes.* San Francisco: Ortho Books, Chevron Chemical Company, 1976.

———. *Gardening Shortcuts.* San Francisco: Ortho Books, Chevron Chemical Company, 1974.

Pavel, Margaret Brandstrom. *Gardening with Color.* San Francisco: Ortho Books, Chevron Chemical Company, 1977.

*Permaculture Quarterly.* 37 Goldsmith St., Maryborough, Victoria, Australia.

Philbrick, Helen, and Gregg, Richard. *Companion Plants and How to Use Them.* Old Greenwich, Conn.: Devin-Adair Company, 1966.

———, and Philbrick, John. *The Bug Book: Harmless Insect Controls.* Charlotte, Vermont: Garden Way Publishing, 1974.

Phillips, Roger. *Trees of North America and Europe.* New York: Random House, 1978.

Riotte, Louise. *Success with Small Food Gardens, Using Special Intensive Methods.* Charlotte, Vermont: Garden Way Publishing, 1977.

Rodale, J. I., editor-in-chief, and the staff of *Organic Gardening and Farming Magazine. The Encyclopedia of Organic Gardening.* Emmaus, Pennsylvania: Rodale Books, 1959.

Rosengarten, Frederick, Jr. *The Book of Spices.* Wynnewood, Pennsylvania: Livingston Publishing Co., 1969.

*Schuler, Stanley. *Gardens Are for Eating.* New York: Macmillan Publishing Company, Collier Books, 1971.

Sedden, George, on gardening; Radecka, Helena, on cookery. *Your Kitchen Garden.* Hawkworth, Swindon, England: Mitchell Beazley, Publishers, and Edenlite, 1975.

Shoemaker, James Sheldon. *Small Fruit Culture.* New York: McGraw-Hill Book Company, 1955.

Simmons, Alan E. *Growing Unusual Fruit.* New York: Walker & Company, 1972.

Smith, J. Russell. *Tree Crops: A Permanent Agriculture.* Old Greenwich, Conn.: Devin-Adair Company, 1950.

Spurr, Joy. *Wild Shrubs: Finding and Growing Your Own.* Seattle: Pacific Search Press, 1978.

Sturtevant, E. L. *Sturtevant's Edible Plants of the World,* ed. U. P. Hedrick, 1919. Reprint. New York: Dover Publications, 1972.

Thompson, Homer C., and Kelly, William C. *Vegetable Crops,* 5th ed. New York: McGraw-Hill Book Company, 1957.

Thompson, Maxine. Class materials on fruits for "Systematics of Fruits and Vegetables" course. Mimeographed. Corvallis, Ore.: Oregon State University, 1977.

Turner, Nancy J. *Food Plants of the British Columbia Indians: Part I/ Coastal Peoples.* Victoria, British Columbia: British Columbia Provincial Museum, 1975.

———. *Food Plants of the British Columbia Indians: Part II/Interior Peoples.* Victoria, B.C.: British Columbia Provincial Museum, 1978.

Tyler, Hamilton. *Gourmet Gardening.* New York: Van Nostrand Reinhold Company, 1972.

*U.S. Department of Agriculture. *Yearbook of Agriculture, 1977: Gardening for Food and Fun.* Washington, D.C.: U.S. Government Printing Office, 1977.

Uphof, J. C. Th. *Dictionary of Economic Plants.* New York: Hafner Publishing Company, 1959.

*Vilmorin-Andrieux, Mm. *The Vegetable Garden,* 1889. Reprint. Palo Alto, Calif.: Jeavons-Leler Press, 1976.

Westwood, Melvin N. *Temperate-Zone Pomology.* San Francisco: W. H. Freeman, 1978.

Wilson, Helen Van Pelt. *The Joy of Geraniums.* New York: William Morrow & Company, 1946.

*Wyman, Donald. *Wyman's Gardening Encyclopedia.* New York: Macmillan Publishing Company, 1971.

——— and Nelson, Peter K., eds. *Handbook on Vines.* Brooklyn Botanic Garden Handbook #14. Brooklyn, New York: Brooklyn Botanic Garden, 1954.

Yashiroda, Ken, and Woodward, Carol H., eds. *Handbook on Japanese Herbs and Their Uses.* Brooklyn Botanic Garden Handbook No. 57. Brooklyn, New York: Brooklyn Botanic Garden, 1954.

Zucker, Isabel. *Flowering Shrubs.* New York: D. Van Nostrand Company, 1966.

# Index

*Abelmoschus esculentus,* 130–131
*Acer macrophyllum,* 192
*Acer rubrum,* 192
*Acer saccharinum,* 192
*Acer saccharum,* 192
*Achillea millefolium,* 199
Acorn squash, 170, 171
*Actinidia arguta,* 226
*Actinidia chinensis,* 226
Adzuki bean, 193
*Agastache foeniculum,* 51–52
*Agastache mexicana,* 52
*Agastache rugosa,* 52
Ailanthus, 39
Akebia, 41, 191
*Akebia lobata,* 191
*Akebia quinata,* 191
*Akebia trifoliata,* 191
Albizia, 39
*Alcea rosea,* 195
Allegheny serviceberry tree, 213
*Allium ampeloprasum,* Porrum Group, 113–115
*Allium cepa,* 131–134
*Allium cepa,* Aggretatum Group, 134
*Allium cepa,* Cepa Group, 188
*Allium cepa,* Proliferum Group, 134
*Allium cepa* var. *viviparum,* 131–134
*Allium cernuum,* 133
*Allium chinense,* 80–82
*Allium fistulosum,* 131–134
*Allium fistulosum* x *allium cepa,* 131–134
*Allium hybridus* var. *erythrostachys,* 51
*Allium odorum,* 80–82
*Allium porrum,* 113–115
*Allium ramosum,* 80–82
*Allium sativum,* 134, 187
*Allium schoenoprasum,* 79–80
*Allium scorodoprasm,* 134
*Allium senescens,* 133
*Allium sphaerocephalum,* 133–134
*Allium tricolor* var. *salicifolius,* 51
*Allium tuberosum,* 80–82
Almond tree, 219
Alpine strawberry, 226–227
*Altaica* (rose), 154
*Althaea officinalis,* 196
*Althaea rosea. See* Hollyhock
*Amaracus dictamnus,* 136
Amaranthus, 49–51
*Amaranthus caudatus,* 49–51
*Amaranthus gangeticus,* 49–51
*Amaranthus hybridus* var. *erythrostachys,* 51
*Amaranthus hypochondriacus.See* Amaranthus
*Amaranthus tricolor* 'Splendens,' 49–51
*Amaranthus tricolor* var. *salicifolius,* 51
*Amelanchier,* 212–213, 223
*Amelanchier alnifolia,* 223
*Amelanchier canadensis,* 212–213
*Amelanchier florida,* 213
*Amelanchier grandiflora,* 213
*Amelanchier laevis,* 213
*Amelanchier stolonifera,* 223
American beech tree, 219
American cress, 186
American elder tree, 212
American fawn lily. *See* Dogtooth violet
American fox grape, 108
American lotus, 196
Anasillo. *See* Sweet-scented marigold
*Anethum graveolens,* 92–93
*Anethum sowa,* 93
Angelica, 193
*Angelica archangelica,* 193
Anise, 185
Anise hyssop, 51–52
Annual flowers, 42
*Anthemis nobilis,* 191
*Anthriscus cerefolium,* 186
Apio, 78
*Apios americana,* 195
*Apios tuberosa. See* Groundnut

*Apium graveolens* var. *dulce,* 77–79
*Apium graveolens* var. *rapaceum,* 77–79
Apple mint, 124
Apple rose, 153
Apple serviceberry tree, 213
Apple tree, 205
Apricot tree, 211
*Arachis hypogaea,* 196
*Araucaria araucana,* 192
*Arbutus unedo,* 192
*Arctium lappa,* 185
*Arctostaphylos uva-ursi,* 191
*Armoracia rusticana,* 187
Arrowhead, 44, 52–54
  Chinese, 53, 54
*Artemisia dracunculus,* 189
Artichoke, 41
  globe, 54–55
  Jerusalem, 36, 185
Asafetida, 193
*Asarum canadense,* 198
Asiatic lotus. *See* Sacred lotus
*Asimina triloba,* 213
Asparagus, 55–57
Asparagus bean, 199
Asparagus lettuce. *See* Celtuce
*Asparagus officinallis,* 55–57
Asparagus pea, 193
*Atriplex hortensis,* 196
Avalanche lily, 195

Baby lima bean, 121
Bamboo, 194
Banana pepper, 145
*Barbarea praecox,* 186
*Barbarea verna,* 186
Barberry, Darwin, 191
*Basella alba,* 196
Basil, 57–58
Bay laurel, 194
Bean, 36
  adzuki, 193
  asparagus, 199
  broad, 186
  bush, 59–60
  dry, 60
  fava, 186
  garbanzo, 186
  goa, 193
  lima, 120–121
  pole, 60–63
  runner, 60–63
  snap, 59–60
  yard-long, 199
Bearberry, 191
Beebalm, 194
Beech, American, tree, 219
Beefsteak plant, 196
Beet, 63–64
Bell pepper, 143, 144–145
Bellflower, 194
*Benincasa cerifera,* 103–105
*Benincasa hispida,* 103–105
*Berberis darwinii,* 191
Bergamot, 194
  lemon, 194
  mint. *See* Orange mint
  wild. *See* Bergamot
Bermuda buttercup, 196
Bermuda sorrel. *See* Sourgrass
Berry (*see also* specific berry)
  buffalo, 223
  oso, 223
  partridge, 193
  vine, 36
*Beta vulgaris,* Cicla Group, 174–175
*Beta vulgaris,* Crassa Group, 63–64
Bibb lettuce, 119
Bird's nest gourd, 106
Black bamboo. *See* Bamboo
Black cherry tree, 209
Black cumin, 195
Black haw, 222
Black huckleberry, 228
Black mulberry tree, 211
Black mustard, 127
Black raspberry, 225
Black salsify, 188
Black walnut tree, 214–215
Blackberry, 224, 225
  thornless, 41, 225
Blackeye pea, 195
Blue-stemmed taro. *See* Violet-stemmed taro
Blueberry, 41, 219–220
  lowbush, 228
Blueberry elder tree, 212
Bok choy, 71–72
Borage, 64–66
*Borago officinalis,* 64–66
Borecole. *See* Kale
Bottle gourd, 106
Box huckleberry, 228
Boysenberry, 225
*Brasenia schreberi,* 198
*Brassica alba,* 126–127
*Brassica arvensis,* 126–127
*Brassica caulorapa,* 187

*Brassica chinensis*, 70–72
*Brassica hirta*, 127
*Brassica japonica*, 125–126
*Brassica juncea*, 126–127
*Brassica kaber*, 127
*Brassica napus*, Napobrassica Group, 188
*Brassica nigra*, 127
*Brassica oleracea*, Acephala Group, 99–100, 111–113, 186
*Brassica oleracea*, Alboglabra, 187
*Brassica oleracea*, Botrytis Group, 185
*Brassica oleracea*, Capitata Group, 67–70
*Brassica oleracea*, Gemmifera Group, 185
*Brassica oleracea*, Gongylodes Group, 187
*Brassica oleracea*, Italica Group, 185
*Brassica pekinensis*, 70–72
*Brassica perviridis*, 126–127
*Brassica rapa*, Chinensis Group, 70–72
*Brassica rapa*, Pekinensis Group, 70–72
*Brassica rapa*, Perviridis Group, 127
*Brassica rapa*, Rapifera Group, 189
Broad bean, 186
Broadleaf evergreens, 13, 32
Broccoli, 185
Brussels sprouts, 185
Buffalo berry, 223
Bulbing onion, 188
Bunching onion, 131–134
Burdock, 185
Burnet, 66–67
Burnet rose, 153
Bush bean, 59–60
Bush lima bean, 120–121
Bush pea, 139–140
*Butomis umbellatus*, 195
Butterhead lettuce, 116–117, 119
Butternut squash, 170–171
Butternut tree, 216

Cabbage, 67–70
  Chinese, 70–72
  flower, 99–100
  ornamental, 99–100
  savoy, 70
Cabbage rose, 153
Cadmium in city vegetable gardens, 10
Calendula, 72–73
Calendula officinalis, 72–73
Caltrops, 184, 194
*Campanula rapunculus*, 194
Canaf, 194
Cane fruits, 224–226
Cantaloupe, 188
Cape gooseberry, 111
Caper, 194
*Capparis spinosa*, 194
*Capsicum annuum*, 142–145
Caraway, 194
*Cardamine pratensis*, 195
Cardoon, 185
Carnation, 74–75
Carosella, 98
Carrot, 75–77
*Carum carvi*, 194
*Carya illinoinensis*, 216
Carya laciniosa, 217
*Carya ovata*, 217
*Castanea crenata*, 218
*Castanea mollissima*, 217–218
Catnip, 191
Cauliflower, 185
Cayenne pepper, 144
Celeriac, 77–79
Celery, 77–79
Celery root, 77–79
*Celtis*, 191
Celtuce, 186
*Centranthus ruber*, 197
*Cercis canadensis*, 197
*Cercis siliquastrum*, 197
*Chaenomeles japonica*, 222–223
*Chaenomeles speciosa*, 222–223
*Chaerophyllum bulbosum*, 198
*Chamaemelum nobile*, 191
Chamomile, 191
Chard. *See* Swiss chard
Chayote, 194
Checkerberry, mountain, 228
*Chenopodium album*, 187
Cherry
  cornelian, 41, 222
  eleagnus, 223
  ground, 109–111
  Korean bush, 222
  Nanking, 222
  sand, 222
  sweet, 41
  western bush, 222
Cherry pepper, 144
Cherry tomato, 179, 180
Cherry tree, 208–209
Chervil, 186
  turnip-rooted, 198
Chickpea, 186
Chicory, 186
Chili pepper, 143–144
Chinese arrowhead, 53, 54

Chinese cabbage, 70–72
Chinese celery cabbage, 70–72
Chinese chestnut tree, 217–218
Chinese chives, 80–82
Chinese flowering chives. *See* Chinese chives
Chinese flowering leek. *See* Chinese chives
Chinese garlic. *See* Chinese chives
Chinese hawthorn tree, 212
Chinese kale, 187
Chinese-lantern plant, 111
Chinese mustard cabbage, 71–72
Chinese parsley. *See* Coriander
Chinese potherb mustard. *See* Mizuna
Chinese preserving melon. *See* Winter melon
Chinese radish. *See* Fall radish
Chinese yam. *See* Yam
*Chiognes hispidulum. See* Oregon wintergreen
Chives, 79–80
  Chinese, 80–82
  garlic, 80–82
Chop suey greens. *See* Chrysanthemum greens
*Chrysanthemum coronarium,* 82–83
Chrysanthemum greens, 82–83
*Chrysanthemum morifolium,* 83
*Chrysanthemum spatiosum,* 82–83
*Cicer arietinum,* 186
*Cichorium endivia,* 96–97
*Cichorium intybus,* 186
Cilantro, 84–85
Cinnamom vine. *See* Yam
*Citrullus lanatus,* 189
*Citrullus vulgaris,* 189
City gardens, 8–10
Clary sage, 158
*Claytonia perfoliata. See* Miner's lettuce
Clematis, 41
Cleome, 42
Clove pink, 74–75
Clove-scented basil, 58
Clove-scented currant, 221
Clover, lucky, 196
Coast strawberry, 227
Collards, 186
*Colocasia antiquorum. See* Taro
*Colocasia esculenta,* 198
Comfrey, 195
Common onion, 188
Common wood sorrel, 196
Common yellow pond lily, 196
Community gardens, 10
Conical pepper, 144–145
Container gardening, 7–8
Cooperative Extension Service, 10, 45, 49
Copper fennel, 97, 98
Coriander, 84–85
*Coriandrum sativum,* 84–85
Corn, 36, 42, 85–88
Corn salad, 186
Cornelian cherry, 41, 222
*Cornus mas,* 222
Corsican mint, 124, 125
*Corylus,* 218
*Corylus avellana,* 218
*Corylus colurna,* 218
*Corylus maxima,* 218
Cos lettuce, 119
Cowberry, 228
Cowpea, 195
Crabapple tree, 210
*Crambe maritima,* 197
Cranberry, 228
*Crataegus,* 212
*Crataegus aestivalis,* 212
*Crataegus coccinoides,* 212
*Crataegus mollis,* 212
*Crataegus pinnatifida* var. *major,* 212
Creeping thyme, 41, 176, 177
Cress
  American, 186
  garden, 186
  upland, 186
Crisphead lettuce, 119
*Crocus sativus,* 197
Crookneck squash, 167
Crown daisy. *See* Chrysanthemum greens
*Cryophytum crystallinum. See* Ice plant
*Cryptotaenia canadensis,* 196
*Cryptotaenia japonica. See* Mitsuba
Cuckoo flower, 195
Cucumber, 36, 88–90
  serpent, 197
*Cucumis anguria,* 90
*Cucumis melo,* 188
*Cucumis sativus,* 88–90
*Cucurbita ficifolia,* 171
*Cucurbita maxima,* 168–171
*Cucurbita mixta,* 168–171
*Cucurbita moschata,* 168–171
*Cucurbita pepo,* 165–167, 168–171
Cumin, 190
  black, 195
*Cuminum cyminum,* 190
Currant, 42, 220–221

*Cydonia oblonga,* 210
*Cynara cardunculus,* 185
*Cynara scolymus,* 54–55

Daikon (fall radish), 147–149
Daisy, crown. *See* Chrysanthemum greens
Damask rose, 153
Dandelion, 186
Darwin barberry, 191
Dasheen, 198
*Daucus carota* var. *sativus,* 75–77
David's peach tree, 209
Day lily, 39, 90–92
DeDeurwaerder, Chuck, 16
Designing food garden, 15–42
  analyze site, 21, 23–28
  analyze use of site, 16–21
  with annual flowers, 42
  combine analyses, 29–37
  determine space, 20–21
  with inedible ornamentals, 38–42
  make scale maps, 25
  position plants, 36–37
  work with textures and plants, 32–35
Devil's claws. *See* Unicorn plant
*Dianthus,* 74–75
*Dianthus carophyllus,* 74–75
*Dianthus plumarius,* 75
Dill, 92–93
*Dioscorea alata,* 198
*Dioscorea batatas,* 198
*Diospyros chinensis. See* Oriental persimmon
*Diospyros kaki,* 210–211
*Diospyros virginiana,* 211
Dipper gourd, 106
Dittany of Crete (marjoram), 136
Doan gwa. *See* Winter melon
Dogtooth violet, 195
Dow gauk. *See* Yard-long bean
Downy hawthorn tree, 212
Dropwort, 195
Dry beans, 60
Duck potato. *See* Arrowhead
Duke cherry tree, 208–209
Dwarf bush nasturtium, 128–130
Dwarf cape gooseberry. *See* Husk tomato
Dwarf dill, 93
Dwarf kale, 111–113
Dwarf pea, 141
Dwarf tomato, 180, 181

East Indian lotus, 196
Edible-podded pea, 141
Edible snake gourd, 197
Eggplant, 94–95
Egyptian onion, 134
*Elaeagnus angustifolia,* 192
*Elaeagnus argentea,* 192
*Elaeagnus commutata,* 192
*Elaeagnus multiflora,* 223
*Elaeagnus umbellata,* 192
Elderberry tree, 212
*Eleocharis dulcis,* 183–184
*Eleocharis tuberosa,* 183
Elephant garlic, 115
Endive, 96–97
English lavender, 196
English thyme, 176–177
English walnut tree, 214
Environmental Protection Agency, 10
*Eruca sativa,* 190
*Eruca vesicaria,* 190
*Erythronium americanum,* 195
*Erythronium dens-canis,* 195
*Erythronium grandiflorum,* 195
*Erythronium oregonum,* 195
Escarole, 96–97
European cranberry, 228
European elder tree, 212
European fawn lily, 195
European pond lily, 197
*Eutrema wasabi,* 198
Evaluating food plants, 43–44, 45–48
Evening primrose, 195
Evergreens, 212, 223
  broadleaf, 13, 32
  in food garden, 38, 39, 40
*Exogonium bracteatum. See* Jicama

*Fagus grandifolia,* 219
Fall radish, 147–149
Fava bean, 186
Fennel, 97–99
Fennel flower, 195
Fenugreek, 190
Fern, 41
Fernleaf geranium, 103
*Ferula foetida,* 193
Fetticus. *See* Corn salad
*Ficus carica,* 211
Field corn, 85–88
Field poppy, 145–147
Fig tree, 211
Filbert tree, 218
*Filipendula hexapetala. See* Dropwort
*Filipendula vulgaris,* 195
Finocchio, 97
Firethorn, 192

Fishpole bamboo. *See* Bamboo
Florence fennel, 97, 98
Floribunda rose,153
Flower cabbage, 99–100
Flower kale, 99–100
Flowering raspberry, 225
Flowering rush, 195
Flowers, annuals, in food garden, 42
*Foeniculum dulce,* 97–99
*Foeniculum vulgare,* 97–99
Food garden
  choosing plants for, 13–14
  city, 8–10
  community, 10–11
  container, 7–8
  designing. *See* Designing food garden
  large, 12–13
  small, 8
  suburban, 12
Food plants
  evaluating, 43–44, 45–48
  to explore, 193–199
  less attractive, 185–190
  with limited uses, 191–193
  sources of, 44–45, 229–232
*Fragaria,* 226–227
*Fragaria chiloensis,* 227
*Fragaria vesca,* 227
*Fragaria virginiana,* 227
French rose, 153
French thyme, 176–177
Fruit
  bushes, 219–224
  cane, 224–226
  ground cover, 226–228
  plants, 200–204
  trees, 11–12, 201–203, 204–213
  vine, 226
Fuchsias, 223–224
*Fuchsia corymbiflora,* 223
*Fuchsia denticulata,* 223
*Fuchsia racemosa,* 223
Fuzzy gourd, 103–105

*Galium odoratum,* 193
Garbanzo bean, 186
Garland chrysanthemum. *See* Chrysanthemum greens
Garlic, 134, 187
  Chinese. *See* Chinese chives
  elephant, 115
  serpent, 134
Garlic chives, 80–82
*Gaultheria hispidula,* 228
*Gaultheria ovatifolia,* 228
*Gaultheria shallon,* 223
*Gaylussacia baccata,* 228
*Gaylussacia brachycera,* 228
*Gaylussacia dumosa,* 228
*Gaylussacia frondosa,* 228
Geranium, scented, 100–103
Gerard, John, 74–75, 173
German chamomile, 191
Gherkin, West Indian, 90
Giant timber bamboo. *See* Bamboo
Gibbons, Euell, 183
Ginger, wild, 198
Ginger geranium, 103
*Gingko biloba,* 219
Gingko tree, 219
Girasole. *See* Jerusalem artichoke
*Glaucum,* 133
Globe artichoke, 54–55
*Glycine max,* 162–164
Goa bean, 193
Golden mint, 125
Gongylodes Group, 187
Gooseberry, 221, 226
Goosefoot. *See* Lambsquarters
Goumi, 223
Gourd, 36, 106
  edible snake, 197
  fuzzy, 103–105
  lagenaria, 105–107
  sponge, 196
  wax. *See* Fuzzy gourd
Gow choy. *See* Chinese chives
Grandiflora rose, 153
Grape, 36, 41, 107–109, 226
  American fox, 108
  muscadine, 108
  Oregon, 223, 228
  parsley, 109
Great burnet, 67
Great-headed garlic. *See* Elephant garlic
Greek oregano, 135
Green cabbage, 67–70
Green gram. *See* Mung bean
Ground cherry, 109–111
Ground covers, 41, 226–228
Groundnut, 195

Hackberry, 191
Hawthorn tree, 212
Hazelnut tree, 218
He shi ko. *See* Bunching onion
Heartnut tree, 216
*Helianthus annuus,* 172–173

*Helianthus giganteus,* 173
*Helianthus tuberosus,* 185
*Hemerocallis aurantiaca,* 91
*Hemerocallis fulva,* 90–92
*Hemerocallis lilioasphodelus,* 91
*Hemerocallis minor,* 91
Herb varieties (*see also* specific herb)
  evaluating, 43–44, 45–48
*Hibiscus cannabinus,* 194
*Hibiscus sabdariffa,* 197
Hickory tree, 217
Highbush blueberry, 219–220
Hinn choy. *See* Amaranthus
*Hippophae rhamnoides,* 192
Hobblebush, 222
Hollyhock, 195
Hon tsai tai, 72
Honewort, 196
Honey locust, 39
Honeydew melon, 188
Honeysuckle, 41
Hop marjoram, 136
Horehound, 191
Horse bean. *See* Fava bean
Horsemint. *See* Bergamot
Horseradish, 187
  Japanese, 198
Hot pepper, 142–145
Huckleberry
  black, 228
  box, 228
Husk cherry. *See* Husk tomato
Husk tomato, 109–111
Hybrid tea rose, 153
Hyssop, 191
  anise, 51–52
  Mexican giant, 52
*Hyssopus officinalis,* 191

Ice plant, 196
Iceberg lettuce,116–117, 119
*Ipomoea batatas,* 197
India mustard. *See* Mustard greens
Indian hemp, 194
Indian plum, 223
Indian sorrel. *See* Roselle
Ivy, 41

Jamaican sorrel, 197
Japanese bunching onion, 131–134
Japanese chestnut tree, 218
Japanese eggplant, 94–95
Japanese greens. *See* Chrysanthemum greens; Mizuna
Japanese horseradish, 198
Japanese parsley, 196
Japanese plum tree, 207–208
Japanese radish. *See* Fall radish
Japanese walnut tree, 216
Jerusalem artichoke, 36, 185
Jicama, 196
Joseph sage, 157, 158
Joseph's coat (amaranthus), 49–51
Judas tree, 197
*Juglans ailanthifolia,* 216
*Juglans ailanthifolia* var. *cordiformis,* 216
*Juglans cinerea,* 216
*Juglans cordiformis.* *See* Heartnut
*Juglans nigra,* 214–215
*Juglans regia,* 214
*Juglans sieboldiana.* *See* Japanese walnut
*Juglans* x *Juglans sieboldiana,* 216
Juniper, prostrate, 41

Kailaan, 187
Kale, 41, 111–113
  Chinese, 187
  flower, 99–100
  ornamental, 99–100
  sea, 197
Kansas hawthorn tree, 212
Katsura tree, 39
Kettle gourd, 106
Kiwi, 226
*Koellia virginiana.* *See* Virginia mountain mint
*Koelruteria,* 39
Kohlrabi, 187
Korean bush cherry, 222
Korean mint, 52
Korean pine tree, 219
Kudzu, 187
Kurrat, 115
Kyona. *See* Mizuna

*Lactuca sativa,* 116–119, 186
Lady's smock, 195
Lagenaria gourd, 105–107
*Lagenaria longissima,* 106
*Lagenaria siceraria,* 105–107
*Lagenaria vulgaris,* 105–107
Lambsquarters, 187
*Laurus nobilis,* 194
*Lavandula angustifolia,* 196
*Lavandula officinalis,* 196
*Lavandula spica* var. *angustifolia.* *See* English lavender

*Lavandula vera. See* English lavender
Lavender, English, 196
Lead content, in city vegetable gardens, 9–10
Leek, 35, 41, 113–115
  Chinese flowering. *See* Chinese chives
Legumes. *See* specific legume
Lemon balm, 115–116
Lemon basil, 58
Lemon bergamot, 194
Lemon day lily, 91
Lemon geranium, 102, 103
Lemon thyme, 177
*Lens culinaris,* 188
Lentil, 188
*Lepidium sativum,* 186
Lettuce, 48, 116–119
  miner's, 196
*Levisticum officinale,* 121–123
*Lilium lancifolium,* 198
*Lilium tigrinum. See* Tiger lily
Lily
  avalanche, 195
  day, 39, 90–92
  European fawn, 195
  lemon day, 91
  pond, 196–197
  tiger, 198
  trout, 195
  yellow, 91
Lima bean, 120–121
Limber pine tree, 219
Lindsey, Steve, 16
Lingonberry, 228
Little-leaf basil, 57–58
Live-forever, 197
Loganberry, 225
Loose-leaf lettuce, 117–119
Lotus
  American, 196
  East Indian, 196
  sacred, 196
*Lotus tetragonolobus. See* Asparagus pea
Lovage, 121–123
Love-in-a-mist, 195
Love-lies-bleeding (amaranthus), 48
Lowbush blueberry, 228
Lucky clover, 196
Luffa, 196
*Luffa acutangula,* 196
*Luffa aegyptiaca,* 196
*Luffa cylindrica. See* Luffa
*Lycopersicon esculentum,* 177–181
*Lycopersicon lycopersicum,* 177–181
*Lycopersicon pimpinellifolium,* 181

Mace, sweet, 196
Mache (corn salad), 186
Magnolia, 39
*Mahonia,* 223
*Mahonia aquifolium,* 223
*Mahonia nervosa,* 223
*Mahonia repens,* 228
*Majorana hortensis,* 135
*Majorana onites,* 135
Malabar spinach, 196
*Malus,* 210
*Malus baccata,* 205, 210
*Malus baccata* var. *mandschurica,* 210
*Malus coronaria,* 210
*Malus pumila,* 205
*Malus sylvestris,* 205
Manchu cherry. *See* Nanking cherry
Mango melon, 188
Mao gwa. *See* Fuzzy gourd
Maples, 192
Marigold, 42
  pot, 72–73
  sweet-scented, 196
Marjoram, 188
  hop, 136
  pot, 135
  sweet, 134, 135, 188
  wild, 134–136
Marrow. *See* Summer squash
*Marrubium vulgare,* 191
Marshmallow, 196
*Martynia proboscidea. See* Unicorn plant
Martynia. *See* Unicorn plant.
*Matricaria chamomilla,* 191
*Matricaria recutita,* 191
May apple, 192
Mayhaw tree, 212
Maypops, 226
*Mays,* 85–88
Medlar, 222
*Melissa officinalis,* 115–116
Melon, 188
  winter, 103–105, 188
*Mentha,* 123–125
*Mentha citrata,* 124
*Mentha requienii,* 125
*Mentha spicata,* 124
*Mentha suaveolens,* 124
*Mentha viridis,* 124
*Mentha* x *gentilis,* 124–125

*Mentha* x *gentilis*, var. *variegata*, 125
*Mentha* x *piperita*, 124
*Mentha* x *piperita* var. *citrata*, 124
*Mesembryanthemum crystallinum*, 196
*Mespilus germanica*, 222
Mexican giant hyssop, 52
Miner's lettuce, 196
Miniature rose, 151–153
Mint, 39, 41, 123–125
  apple, 124
  Corsican, 124, 125
  golden, 125
  Korean, 52
  orange, 124
  peppermint, 124
  pineapple, 124
  red, 124–125
  spearmint, 124
  Virginia mountain, 198
*Mitchella repens*, 193
Mitsuba, 196
Mizuna, 125–126
*Monarda citriodora*, 194
*Monarda didyma*, 194
*Monarda fistulosa*, 194
Monkey-puzzle, 192
*Montia perfoliata*, 196
*Morus nigra*, 211
Mother-of-thyme, 177
Mountain ash, 192
Mountain checkerberry, 228
Mulberry tree, 211
Mung bean, 188
Muscadine grape, 108
Muskmelon, 188
Mustard, 127
  spinach, 127
Mustard greens, 41, 126–127
Mustard seed, 127
*Myrrhis odorata*, 197

Nanking cherry, 222
Nannyberry, 222
Napobrassica Group, 188
Nasturtium, 128–130
*Nasturtium officinale*, 198
Native American persimmon tree, 211
*Nelumbo nucifera*, 196
*Nelumbo speciosa*. *See* Sacred lotus
*Nepeta cataria*, 191
New Zealand gooseberry, 226
New Zealand spinach, 164–165
*Nigella damascena*, 195
*Nigella sativa*, 195
Nitschke, Robert, 109
Nodding onion, 133
*Nuphar advena*, 196
*Nuphar luteum*, 197
*Nuphar polysepalum*, 197
Nurseries, 229–232
Nut trees, 213–219
  on city streets, 11–12
Nutmeg geranium, 102

Oak leaf lettuce, 118–119
Oaks, 192
*Ocimum basilicum*, 57–58
*Ocimum basilicum* 'Minimum,' 57–58
*Ocimum minimum*, 57–58
*Oemleria cerasiformis*, 223
*Oenanthe fistulosa*, 123
*Oenothera biennis*, 195
Okra, 130–131
Onion
  bulbing, 188
  bunching, 131–134
  common, 188
  Egyptian, 134
  nodding, 133
  top, 134
  tree, 134
  Welsh, 131–134
Opium, 146–147
Orach, 196
Oregano, 134–136
Oregon grape, 223, 228
Oregon wintergreen, 228
Oriental persimmon tree, 210–211
Oriental poppy, 41
*Origanum dictamnus*, 136
*Origanum hera cleoticum*, 135
*Origanum majorana*, 135, 188
*Origanum onites*, 135
*Origanum vulgare*, 134–136
*Origanum vulgare* 'Viride,' 135
Ornamental cabbage, 99
Ornamental kale, 99
Ornamentals, inedible, 38–40
Orpine, 197
Oso berry, 223
Oswego tea. *See* Beebalm
Oxalis, 196
  rosette, 196
*Oxalis acetosella*, 196
*Oxalis cernua*. *See* Oxalis
*Oxalis deppei*, 196
*Oxalis pescaprae*, 196

Oyster plant, 188

*Pachyrhizus tuberosus,* 196
Pak choy, 71
*Papaver rhoeas,* 145
*Papaver somniferum,* 146
Parsley, 41, 136–138
  Japanese, 196
Parsnip, 138–139
Partridge berry, 193
*Passiflora caerulea,* 226
*Passiflora incarnata,* 226
Passion flower, 226
*Pastinaca sativa,* 138–139
Pattypan squash, 167
Pawpaw tree, 213
Pea, 139–141
  blackeye, 195
Peach tree, 209
Peanut, 196
Pear tree, 206
Pecan tree, 216
*Pelargonium,* 100–103
*Pelargonium crispum,* 102
*Pelargonium denticulatum,* 103
*Pelargonium graveolens,* 102
*Pelargonium tomentosum,* 102
*Pelargonium* x *fragrans,* 102
*Pelargonium* x *nervosum,* 102
Pepper, 142–145,
Peppermint, 124,
Peppermint geranium, 102
Perilla, 196
*Perilla frutescens,* 196
*Perilla laciniata. See* Perilla
*Perilla nankinensis. See* Perilla
Persian walnut tree, 214
Persimmon, 41
Persimmon tree
  native American, 211
  Oriental, 210–211
Pesticides, 45
Pe tsai, 70
*Petroselinum crispum,* 136–138
*Phaseolus angularis,* 193
*Phaseolus aureus,* 188
*Phaseolus coccineus,* 60–63
*Phaseolus lunatus,* 120–121
*Phaseolus* var. *vulgaris humilis,* 59–60
*Phaseolus vulgaris,* 60–63
*Phyllostachys aurea,* 194
*Phyllostachys aureosulcata,* 194
*Phyllostachys bambusoides,* 194
*Phyllostachys dulcis,* 194
*Phyllostachys meyeri,* 194
*Phyllostachys nigra,* 194
*Phyllostachys nuda,* 194
*Phyllostachys viridiglaucescens,* 194
*Phyllostachys viridis,* 194
*Phyllostachys vivax,* 194
*Phylloxera,* 108
*Physalis alkekengi,* 111
*Physalis ixocarpa,* 111
*Physalis peruviana,* 111
*Physalis pruinosa,* 109–111
*Pimpinella anisum,* 185
Pine-scented geranium, 103
Pine tree, 219
*Pinea,* 219
*Pinea cembra,* 219
*Pinea cembroides,* 219
*Pinea coulteri,* 219
*Pinea flexilis,* 219
*Pinea koraiensis,* 219
*Pinea lambertiana,* 219
*Pinea sabiniana,* 219
*Pinea torreyana,* 219
*Pinea wallichiana,* 219
Pineapple mint, 124
Pineapple sage, 157
*Pinus excelsa. See* Himalayan pine
*Pinus griffithii. See* Himalayan pine
Pistachio tree, 219
*Pistacia vera,* 219
*Pisum sativum,* 139–141
Planting, 49
Plum, Indian, 223
Plum tree, 13, 41, 207–208
Pole bean, 60–63
*Poncirus trifoliata,* 193
Pond lily, 196–197
Popcorn, 85–87
Poppy
  field, 145–147
  Oriental, 41
  Shirley, 145–147
*Portulaca oleracea,* 188
Pot marigold, 72–73
Pot marjoram, 135
Potato, 188
  sweet, 197
Potato lima bean, 121
Potentilla, creeping, 41
*Poterium officinale,* 66
*Poterium sanguisorba,* 66–67
Primrose malanga. *See* Yautia
Prince's feather, 51
*Proboscidea jussieui. See* Unicorn plant

*Proboscidea louisianica,* 198
Proboscis flower. *See* Unicorn plant
Provence rose, 153
*Prunus americana,* 208
*Prunus amygdalus. See* Almond
*Prunus armeniaca,* 211
*Prunus avium,* 208
*Prunus avium* x *prunus cerasus. See* Duke cherries
*Prunus cerasifera,* 207
*Prunus cerasus,* 208
*Prunus communis. See* Almond
*Prunus davidiana,* 209
*Prunus domestica,* 207
*Prunus dulcis* var. *dulcis,* 219
*Prunus insititia,* 207
*Prunus japonica,* 222
*Prunus maritima,* 222
*Prunus munsoniana,* 208
*Prunus nigra,* 208
*Prunus persica,* 209
*Prunus salicina,* 207–208
*Prunus serotina,* 209
*Prunus tomentosa,* 222
*Prunus* x *effusus,* 208
*Psophocarpus tetragonolobus,* 193
*Pueraria lobata,* 187
*Pueraria thunbergiana. See* Kudzu
Pumpkin, 168–171
Purple raspberry, 225
Purslane, 188
  winter, 196
*Pycnanthemum virginianum,* 198
Pyracantha, 192
*Pyracantha,* 192
*Pyracantha angustifolia,* 192
*Pyracantha coccinea,* 192
*Pyrus calleryana,* 206
*Pyrus communis,* 206
*Pyrus pyrifolia,* 206
*Pyrus ussuriensis,* 206

*Quercus,* 192
Quince, 222–223
Quince tree, 210

Rabbit-eye blueberry, 220
Radish
  fall, 147–149
  rat-tailed, 149
  spring, 190
Raisin, wild, 222
Rampion, 194
*Raphanus sativus,* 147–149, 190
*Raphanus sativus 'Caudatus,'* 149
Raspberry, 224–225
Rat-tailed radish, 149
Red-flowered currant, 221
Red-leaf lettuce, 118
Red mint, 124–125
Red mulberry tree, 211
Red okra, 131
Red raspberry, 224
Red valerian, 197
*Rheum australe,* 151
*Rheum nobile,* 151
*Rheum palmatum,* 151
*Rheum rhabarbarum,* 149–151
*Rheum ribes,* 151
*Rheum tataricum,* 151
Rhubarb, 149–151
*Rhus copallina,* 192
*Rhus glabra,* 192
*Rhus trilobata,* 192
*Rhus typhina,* 192
*Ribes grossularia. see* Gooseberry
*Ribes hirtellum,* 121
*Ribes odoratum,* 221
*Ribes sanguineum,* 221
*Ribes sativum,* 220–221
*Ribes uva-crispa,* 221
Rocambole, 134
Rocket, 190
Romaine lettuce, 119
Roman chamomile, 191
Roquette, 190
*Rosa,* 151–154
*Rosa canina,* 154
*Rosa centifolia,* 153
*Rosa cinnamomea,* 154
*Rosa damascena,* 153
*Rosa eglanteria,* 154
*Rosa gallica,* 153
*Rosa nutkana,* 154
*Rosa pimpinellifolia,* 153
*Rosa pomifera,* 153
*Rosa rugosa,* 153, 223
*Rosa spinosissima,* 153
*Rosa villosa,* 153, 223
Rose, 41, 153, 223
Rose geranium, 102
Roselle, 197
Rosemary, 154–156
Rosette oxalis, 196
*Rosmarinus officinalis,* 154–156
*Rubus argutus,* 225
*Rubus deliciosus,* 226
*Rubus idaeus,* 224

*Rubus illecebrosus,* 228
*Rubus laciniatus,* 225
*Rubus occidentalis,* 225
*Rubus odoratus,* 225
*Rubus parviflorus,* 225–226
*Rubus phoenicolasius,* 226
*Rubus procerus,* 225
*Rubus spectabilis,* 226
*Rubus ursinus,* 225
*Rubus* x *neglectus,* 225
Rugosa rose, 153
Rum cherry tree, 209
*Rumex acetosa,* 189
Runner bean, 60–63
Russet buffalo berry, 223
Russian caraway. *See* Fennel flower
Russian olive, 39, 192
Rutabaga, 188

Sacred basil, 58
Saffron, 197
Sage, 156–158
*Sagittaria latifolia,* 53
*Sagittaria montevidensis,* 53
*Sagittaria sagittifolia,* 53
Saint Johnswort, 41
Salal, 223
Salsify, 188
  black, 188
*Salvia elegans,* 157
*Salvia horminum,* 158
*Salvia officinalis,* 156–158
*Salvia rutilans,* 157
*Salvia sclarea,* 158
*Salvia varidis,* 157, 158
*Sambucus,* 212
*Sambucus caerulea,* 212
*Sambucus canadensis,* 212
*Sambucus nigra,* 212
Sand cherry, 222
Sand leek. *see* Rocambole
*Sanguisorba minor,* 66
Sassafras, 39
*Satureja hortensis,* 158–159
*Satureja montana,* 159–160
Savory
  summer, 158–159
  winter, 159–160
Scallop squash, 166, 167
Scented geranium, 100–103
Scorzonera, 188
*Scorzonera hispanica,* 188
Scotch lovage, 123
Scotch rose, 153
*Scoticum,* 123
Sea buckthorn, 192
Sea fig. *See* Ice plant
Sea kale, 197
Sea lovage, 123
Sea tomato, 153
Seakale beet. *See* Swiss chard
*Sechium edule,* 194
Sedum, 197
*Sedum album,* 197
*Sedum reflexum,* 197
*Sedum rosea,* 197
*Sedum telephium,* 197
Seed companies, 229–232
*Semiaurundinaria fastuosa,* 194
Serpent cucumber, 197
Serpent garlic, 134
Serviceberry, 223
Serviceberry tree, 212–213
Sesame, 160–162
*Sesamum indicum,* 160–162
Shadblow serviceberry tree, 212–213
Shadbush tree, 212–213
Shagbark hickory tree, 217
Shallot, 134
Shellbark hickory tree, 217
*Shepherdia argentea,* 223
*Shepherdia canadensis,* 223
Shirley poppy, 145–147
Shungiku, 82–83
Silverberry. *See* Russian olive
Simonds, John Ormsbee, *Landscape Architecture: The Shaping of Man's Natural Environment,* 21, 23
*Sium sisarum,* 197
Skirret, 197
Snap bean, 59–60
Snow pea, 141
Snowberry, 228
Soapberry, 223
Soil and Health Society, 10
Soil turning, 36
*Solanum melongena,* 94
*Solanum tuberosum,* 188
Sophora, 39
*Sorbus aucuparia,* 192
Sorghum, 36, 197
*Sorghum bicolor,* 197
*Sorghum vulgare,* 197
Sorrel, 189
  wood, 196
Sour cherry tree, 208–209

Sources of food plants, 44–45, 229–232
Sourgrass, 196
Soybean, 41, 162–164
Spanish radish. *See* Fall radish
Spearmint, 124
Spinach, 190
  malabar, 196
  New Zealand, 164–165
Spinach mustard, 127
*Spinacia oleracea,* 190
Sponge gourd, 196
Spring radish, 190
Squash, 36
  acorn, 170, 171
  butternut, 170–171
  crookneck, 167
  pattypan, 167
  scallop, 166, 167
  summer, 165–167
  winter, 168–171
  yellow, 166–167
  zucchini, 165–167
Stagbush, 222
Strawberry, 36, 41, 226–227
  alpine, 226–227
  bush, 192
  coast, 227
  wild American, 227
Strawberry-raspberry, 228
Strawberry tomato, 109–111
Styrax, 39
Suburban gardening, 12
Sugar pea, 141
Sumac, 192
Summer savory, 158–159
Summer squash, 165–167
Sunchoke. *See* Jerusalem artichoke
Sunflower, 36, 172–173
Swede, 188
Sweet basil, 58
Sweet cherry, 41
Sweet cherry tree, 208–209
Sweet cicely, 197
Sweet corn, 85–88
Sweet fennel, 97–99
Sweet marjoram, 135, 188
Sweet pea, 141
Sweet pepper, 143
Sweet potato, 197
Sweet-scented marigold, 196
Sweet-seeded pine tree, 219
Sweet violet, 181–183
Sweet woodruff, 193
Sweetshoot bamboo. *See* Bamboo
Swiss chard, 174–175
Swiss stone pine tree, 219
*Symphytum officinale,* 195

*Tagetes lucida,* 196
Tamarack, 39
*Taraxacum officinale,* 186
Taro, 198
  violet-stemmed, 199
Tarragon, 189
Tarragon-scented basil, 58
Ten-month yam. *See* Yam
*Tetragonia expansa,* 164–165
*Tetragonia tetragonioides,* 164–165
Thimbleberry, 225–226
Thornless blackberry, 41
Thyme, 41, 175–177
*Thymus,* 175–177
*Thymus caespititius,* 177
*Thymus herba-barona,* 177
*Thymus serpyllum,* 177
*Thymus vulgaris,* 176–177
Tiger lily, 198
Tomatillo or jamberry, 111
Tomato, 36, 177–181
  cherry, 179, 180
  dwarf, 180, 181
  husk, 109–111
  paste, 180
  strawberry, 109–111
  wild, 181
Top onion, 134
Tragopogon porrifolius, 188
Trapa natans, 184, 194
Tree basil, 58
Tree onion, 134
Trees
  fruit, 201–202, 204–213
  nut, 213–219
*Trichosanthes anguina,* 197
Tricolor basil, 58
Trifoliate orange, 193
*Trigonella foenum-graecum,* 190
*Tropaeolum majus,* 128–130
*Tropaeolum minus,* 128–130
*Tropaeolum tuberosum,* 130
Trout lily, 195
Tuberous nasturtium, 130
*Turkestaniana,* 158
Turnip, 189
Turnip-rooted chervil, 198
Twinberry, 193

Unicorn plant, 198

*Vaccinium angustifolium,* 228
*Vaccinium ashei,* 220
*Vaccinium corymbosum,* 219–220
*Vaccinium macrocarpon,* 228
*Vaccinium myrtilloides,* 228
*Vaccinium ovatum,* 220
*Vaccinium oxycoccos,* 228
*Vaccinium pensylvanicum. See* Lowbush blueberry
*Vaccinium vacillans,* 228
*Vaccinium vitis-idaea,* 228
*Valerianella locusta,* 186
*Viburnum alnifolium,* 222
*Viburnum lantanoides. See* Hobblebush
*Viburnum lentago,* 222
*Viburnum opulus,* 222
*Viburnum prunifolium,* 222
*Viburnum trilobum,* 222
*Vica faba,* 186
*Vigna angularis,* 193
*Vigna radiata,* 188
*Vigna sesquipedalis. See* Yard-long bean
*Vigna sinensis. See* Cowpea
*Vigna unguiculata,* 195, 199
Vines, in food garden, 40–41, 226
*Viola odorata,* 181–183
*Viola palmata,* 182–183
*Viola papilionacea,* 183
*Viola sororia,* 183
Violet, 41, 181–183, 196
  dogtooth, 195
Violet-stemmed taro, 199
Virginia mountain mint, 198
*Vitis,* 107–109
*Vitis aestivalis,* 109
*Vitis coignetiae,* 109
*Vitis doaniana,* 109
*Vitis labrusca,* 108, 109
*Vitis labrusca* x *vitis vinifera,* 108, 109
*Vitis rotundifolia,* 108, 109
*Vitis vinifera,* 108

Walnut tree, 214–216
Wapato (arrowhead), 53
Wasabi, 198
*Wasabi japonica,* 198
Water chestnut, 183–184, 194
Water lovage, 123
Watercress, 198
Watermelon, 189
Watershield, 198
Wax gourd. *See* Fuzzy gourd
Welsh onion, 131–134
West Indian gherkin, 90
Western bush cherry, 222
Western pond lily, 197
White-flowered gourd. *See* Lagenaria gourd
White mulberry tree, 211
White mustard, 127
White yam. *See* Yam
Wild American strawberry, 227
Wild basil. *See* Virginia mountain mint
Wild bergamot. *See* Bergamot
Wild ginger, 198
Wild marjoram, 134–136
Wild mustard, 127
Wild raisin, 222
Wineberry, 226
Winter melon, 103–105
Winter purslane, 196
Winter radish. *See* Fall radish
Winter savory, 159–160
Winter squash, 168–171
Winter tarragon. *See* Sweet-scented marigold
Wintergreen, Oregon, 228
Wong bok, 70, 71
Wood sorrel, 196
Woodruff. *See* Sweet woodruff
Wooly mint. *See* Apple mint
Wooly mother-of-thyme, 177
Wormgrass, 197

*Xanthosoma sagittifolium,* 199
*Xanthosoma violaceum,* 199

Yam, 198
Yard-long bean, 199
Yarrow, 199
Yautia, 199
  malanga. *See* Yautia
  yellow, 199
Yellow adder's tongue. *See* Dogtooth violet
Yellow lily, 91
Yellow squash, 166, 167
Yellow yautia, 199
Yellowgroove bamboo. *See* Bamboo
Ysano, 130

*Zea,* 85–88
Zelkova, 39
Zinnia, 35
Zucchini, 165–167